Neural Network Perspectives on Cognition and Adaptive Robotics

Related titles published by
Institute of Physics Publishing

Neural Computing: An Introduction
R Beale and T Jackson

Handbook of Neural Computation
Edited by E Fiesler and R Beale

Neural Network Analysis, Architectures and Applications
Edited by A Browne

Neural Network Perspectives on Cognition and Adaptive Robotics

Edited by

Antony Browne

Nene College, UK

Institute of Physics Publishing
Bristol and Philadelphia

British Library Cataloguing-in-Publication Data

A catalogue record for this book is available from the British Library.

ISBN 0 7503 0455 3

Library of Congress Cataloging-in-Publication Data are available

Published by Institute of Physics Publishing, wholly owned by The Institute of Physics, London

Institute of Physics Publishing, Dirac House, Temple Back, Bristol BS1 6BE, UK

US Editorial Office: Institute of Physics Publishing, The Public Ledger Building, Suite 1035, 150 South Independence Mall West, Philadelphia, PA 19106, USA

Typeset in TEX using the IOP Bookmaker Macros
Printed in the UK by J W Arrowsmith Ltd, Bristol

To Jayne and Liam

Contents

Preface   xiii

PART 1
Representation   1

1   Challenges for Neural Computing   **3**
   1.1   Introduction   3
   1.2   Two schools of thought for intelligent systems   3
       1.2.1   The symbolic AI school   4
       1.2.2   The connectionist school   6
   1.3   Connectionist representations   6
       1.3.1   Localist representations   6
   1.4   Distributed representations   7
       1.4.1   Microfeatures   7
       1.4.2   Coarse coding   7
       1.4.3   Extended and superposed representations   7
   1.5   Dynamic versus static representations   9
   1.6   Spiking neurons   10
   1.7   Challenges posed to connectionism by symbolic AI   10
       1.7.1   Structural systematicity   11
       1.7.2   Macro-systematicity   12
       1.7.3   Variable binding   13
       1.7.4   Performance of inference   13
       1.7.5   Limits on the productivity and width representations   14
       1.7.6   Performing dynamic inferencing   15
       1.7.7   Parallelism at the knowledge level   15
       1.7.8   Performing meta-reasoning   16
   1.8   Other criticisms of connectionism   16
       1.8.1   Biological plausibility   16
       1.8.2   Network scalability and training times   17
   1.9   Conclusion   18

2   Representing Structure and Structured Representations in Connectionist Networks   **20**

2.1   Introduction   20
2.2   Representations   21
    2.2.1   Representing structure in a PSS   22
    2.2.2   Structured representations in a PSS   23
    2.2.3   Structure sensitivity in a PSS   24
    2.2.4   Representing structure in connectionist networks   24
    2.2.5   Structured representations in a connectionist network   33
    2.2.6   Structure sensitive processes in a connectionist network   35
2.3   A practical demonstration   36
    2.3.1   The environment   36
    2.3.2   Performance on RAAM-generated representations   37
    2.3.3   Detecting structured representations in RAAM   40
    2.3.4   Explaining the presence of structure with
            decision hyperplanes   43
    2.3.5   Explaining the performance of the classifiers   46
2.4   Discussion and conclusions   49

3   Chaos, Dynamics and Computational Power in Biologically Plausible Neural Networks   **51**

3.1   Introduction   51
3.2   Symbolic systems   51
3.3   The problem of knowledge explosion in symbol systems   52
3.4   Neural networks   53
3.5   Representing rules and relations in neural networks   53
3.6   Time and binding   54
3.7   Computational power and network dynamics   54
3.8   Experiments on network dynamics and computational power   57
3.9   A simple, biologically plausible, neural-network model   57
3.10 Measuring the intrinsic computation in the network   59
3.11 The Crutchfield and Young algorithm   60
3.12 Symbolic reconstructions of cyclic and chaotic dynamics   61
3.13 Applying Crutchfield and Young to the neural network   62
3.14 Discussion   69

4   Information-Theoretic Approaches to Neural Network Learning   **72**

4.1   Introduction   72
4.2   Information theory   73
    4.2.1   Entropy and information   73
    4.2.2   Infomax and information loss   76
4.3   Principal-components analysis   77
    4.3.1   The linear neuron   78
    4.3.2   The Oja principal component finder   79

	4.3.3	Reconstruction error	80
	4.3.4	The scaling problem	81
	4.3.5	Information maximization	81
	4.3.6	Multi-dimensional PCA	83
	4.3.7	Uncorrelated outputs	83
4.4	Lateral inhibition and anti-Hebbian learning		84
	4.4.1	Decorrelating algorithms	84
	4.4.2	Optimizing information with lateral inhibition	85
	4.4.3	Introducing more realistic constraints	87
4.5	I-Max: maximizing mutual information between output units		87
4.6	Conclusions		89

PART 2
Cognitive Modelling   **91**

5   Exploring Different Approaches towards Everyday Commonsense Reasoning   93

5.1	Introduction		93
5.2	Rule-based reasoning		95
	5.2.1	Rules in general	95
	5.2.2	Formal logics	97
	5.2.3	Extensions to formal logics	98
	5.2.4	Probabilistic approaches	98
	5.2.5	Fuzzy logic	99
	5.2.6	Fuzzy evidential logic	99
	5.2.7	Representation and reasoning with rules	104
5.3	Case-based reasoning		105
5.4	Connectionism		108
	5.4.1	High-level connectionist models	109
	5.4.2	Connectionism and rules	110
	5.4.3	Connectionism and cases	110
	5.4.4	Four criteria in integrating rules and similarities	111
5.5	Conclusions		118

6   Natural Language Processing with Subsymbolic Neural Networks   120

6.1	Introduction		120
6.2	Subsymbolic representations		121
	6.2.1	Properties of subsymbolic representations	121
	6.2.2	Example: subsymbolic representations in sentence case-role assignment	123
6.3	Modeling human language processing		128
	6.3.1	Symbols versus soft constraints	129
	6.3.2	Example: case-role assignment of embedded clauses	130
6.4	Overview of subsymbolic natural language processing		135

6.5	Future challenges	137
6.6	Conclusion	138

7 The Relational Mind — **140**

7.1	Introduction	140
7.2	Problems of consciousness	141
7.3	The relational mind model	141
7.4	Global control	142
7.5	Backward referral in time	144
7.6	Subliminal effects	146
7.7	Return to consciousness	148
7.8	Personality and self	149
7.9	What is it like to be?	150
7.10	The PSYCHE project	151
7.11	Conclusions	152

8 Neuroconsciousness: a Fundamental Postulate — **153**

8.1	Introduction	153
8.2	The fundamental postulate	153
8.3	Defining my own consciousness	154
8.4	The consciousness of others	155
8.5	The power of modelling	156
8.6	The danger of inappropriate modelling	158
8.7	Why neurons?	159
8.8	The mind's eye	162
8.9	Learning and remembering	163
8.10	In a nutshell . . .	165

PART 3
Adaptive Robotics — **167**

9 The Neural Mind and the Robot — **169**

9.1	Introduction	169
9.2	The environmental puppeteer	171
	9.2.1 $Beast_1$	171
	9.2.2 Describing behavioral systems	173
	9.2.3 An artificial neural network brain	176
	9.2.4 Adaptivity: the beast and the hairdryer	177
9.3	Robot training methods	178
	9.3.1 Hebbian learning	178
	9.3.2 Backpropagation learning	179
	9.3.3 Reinforcement learning	182
	9.3.4 Evolutionary learning	184
	9.3.5 Beyond reactive control	186
	9.3.6 A summary of neural networks in self-learning robots	188

9.4   Back to the future   188
        9.4.1   Piaget and the emergence of intelligence   190
        9.4.2   A new representationalism?   193
9.5   Conclusions   194

10 Teaching a Robot to See How it Moves   **195**
10.1   Introduction   195
10.2   The components   197
        10.2.1 Robotics   197
        10.2.2 Vision   198
        10.2.3 Control systems   202
10.3   Using visual feedback in robot control   206
        10.3.1 An experimental setup I   206
        10.3.2 An experimental setup II   211

11 Designing a Nervous System for an Adaptive Mobile Robot   **220**
11.1   Introduction   220
11.2   Modelling a nervous system   222
11.3   Neurophysiology of learning   224
        11.3.1 Non-associative learning   224
        11.3.2 Associative learning   226
11.4   Central pattern generators   227
11.5   The computational neuroscience paradigm   229
11.6   *Hi-NOON* neural simulator   231
        11.6.1 *Network* objects   233
        11.6.2 *Neuron* objects   233
        11.6.3 *Synapse* objects   234
        11.6.4 State-system algorithm   234
        11.6.5 Network learning   234
11.7   Simulation of behaviour with *Hi-NOON*   235
11.8   Adaptive robot behaviour   237
        11.8.1 The five-neuron 'trick'   238
        11.8.2 A more complex adaptive vehicle   242
11.9   Discussion and conclusions   248

Bibliography   **251**

Index   **269**

Preface

Neural networks have been used extensively to model human cognition. Human reasoning and language are two areas where, so far, no computational system has reached the level of human performance. There is much interest in neural network models as they provide an approach to the modelling of cognition different from that of traditional symbolic artificial intelligence (AI). Perhaps because of the success of neural network systems in modelling some aspects of cognition and human language processing, many questions have been raised by the symbolic AI community as to the representational power of connectionist systems. Some of these questions are discussed in depth in this book.

Most of us are familiar with images of robots, and often these robots can be seen to be performing useful tasks. However, most of the robots we are familiar with have an important limitation. They must be pre-programmed for a fixed environment and a fixed set of tasks. If their environment changes they fail ungracefully. Adaptive robots using neural computing techniques are able to change their behaviour as their environment changes.

This book is designed to give an overview of some of the many perspectives that neural computing gives on cognition and autonomous robotics. It is not written as an introductory textbook; it is assumed that the reader has some previous knowledge of neural networks and an understanding of their basic mechanisms. The book is divided into several parts:

- *Representation.* This part delves into questions on the representational power of neural networks and discusses the use of information theory in neural computing.
- *Cognitive modelling.* The use of neural networks for cognitive modelling is discussed, including both the modelling of human reasoning and the implications of neurophysiological data.
- *Adaptive robotics.* Robots which can adapt to their environment are described, together with a discussion on biologically realistic learning mechanisms.

Many chapters in the book are cross-referenced to a companion volume *Neural Network Analysis, Architectures and Algorithms* (1997, Institute of

Physics Publishing), which can be seen as complementary to this volume as it contains chapters on the following areas:

- *Understanding and simplifying networks.*   Methods for extracting information about what a trained neural network has learned are outlined, together with a method for simplifying network architectures based on information theory.
- *Novel architectures and algorithms.* Two novel hardware implementations of neural networks are described, together with a discussion of fast training algorithms for feed-forward network architectures.
- *Applications.* Some applications of neural networks in the diverse fields of control (including neuro-fuzzy control), data compression and target identification are discussed.

Hopefully the following chapters will clarify some of the many fields in which neural computing is expanding in its attempts to model human cognition more accurately and produce flexible and adaptive robotic systems. Perhaps in the (possibly very distant) future these two areas will meet, and we will have robotic systems with human-like capabilities of thought and speech.

No textbook can ever hope to give a comprehensive review of the myriad directions in which current research is headed. However, this volume attempts to give a flavour of some of the most promising areas.

A Browne
July 1997

PART 1

REPRESENTATION

Now that the hysteria has died down about neural computing being a magical solution, both to the problems faced by cognitive scientists in modelling human cognition and to the problems faced by engineers in developing intelligent solutions to solve practical problems, it is time to take stock of the situation. Several questions can be asked:

How powerful are neural networks? What sort of things can they represent and what operations can they perform? Are there limits to the sort of operations they can perform and the types of structures they can represent? How do they compare to symbolic computational systems?

Because of the recent popularity of neural computing and the perhaps extravagant claims put forward by some of its proponents, the field has come under fire from many directions. This part of the book concentrates on aspects of the representational power of neural networks and how some of the challenges regarding this representational power have been answered. A formal analysis of what is happening inside neural network systems, using information theory, is also discussed.

Chapter 1

Challenges for Neural Computing

Antony Browne
School of Information Systems, Nene College, UK
antony.browne@nene.ac.uk

1.1 Introduction

In recent years neural networks have been touted as the miracle solution, both to
the problems that cognitive scientists face in trying to build models of cognition
and to the practical problems that engineers face in trying to develop intelligent
solutions for commerce and industry. However, neural networks have also been
subject to much criticism. Some of these criticisms, together with the attempts
that scientists and engineers working with neural networks have made to answer
them, are given below.

1.2 Two schools of thought for intelligent systems

A substantial proportion of artificial intelligence (AI) research, and the
application of the technology produced from this research, is based on the
assumption that all the important aspects of human cognition, and all the
useful tasks which humans require their machines to do, may (at some
time) be described or carried out by a computational model. This view
(computationalism) follows on from the belief that any aspect of cognition or
practical task can be modelled computationally (the Church–Turing thesis). This
can be contrasted with the approach of some theorists who hold that certain
properties of the mind (and hence some properties that we may wish to capture
in practical intelligent systems) cannot be captured algorithmically [228]. Many
researchers in cognitive science believe that implementation details are irrelevant
for cognition, as do many engineers who implement the practical intelligent
systems currently used in commerce and industry. As a result of this some
of those involved with these systems have not tried to obtain inspiration from

the physical structure of the brain, but have instead concentrated on modelling the abstract structure of the mind and the practical application of the associated technology. In recent years there has been much disagreement as to how to produce systems with the desired properties and research in intelligent systems can be seen to have split into two schools. These are the classical (in so far as they have been researching in the field of AI for the longest time) symbolic AI school and the connectionist school. It is a matter of some debate as to which school actually has the most powerful model and members of the connectionist school implementing their models with neural networks are currently attempting to face up to the criticisms levelled at their models by the symbolic AI camp. One could argue that since both neural networks and classical symbolic systems are universal Turing machines [111], at one level of abstraction there is no distinction between them. However, in the nature of the representations they use, in their observed performance given different tasks, and in their adequacy for the modelling of intelligent systems, there are many differences between these two schools of thought.

1.2.1   The symbolic AI school

The classical symbolic school states that the correct level at which to model the mind is that of the symbol, an entity in a computer program that is taken to refer to an entity in the real world. The main assumptions on which the symbolic AI paradigm rests were first explicitly stated under the 'physical symbol system hypothesis' [210] which states that 'a physical symbol system has the necessary and sufficient means for general intelligent action'. In this definition:

- Necessary means that any physical system that exhibits general intelligence will be an instance of a physical symbol system (PSS).
- Sufficient means that any physical symbol system can be organized further to exhibit general intelligent action.
- General intelligent action means the same scope of intelligence seen in humans.

This implies that in real situations behaviour can occur that is both appropriate to the needs of the system and adaptive to the demands of the external environment (within some physical limits imposed by processing speed and memory requirements). This hypothesis states that an entity can only be intelligent if it instantiates a PSS. Symbols correspond to unitary concepts and are taken to represent objects, events, relations between objects and relations between events. In this way the word symbol comes to represent a concept or entity we can put a meaningful label on such as **'apple'** or **'dog'**. At any one time a symbol represents a single entity or concept. Symbols are atomic, they may combine to form symbol structures, but individual symbols may not be broken down. An obvious example of such symbols occurs in programming languages such as Prolog, where atoms such as **'man'**, **'fred'**, or '**tuesday**' are symbols and uninstantiated (or unbound) variables are symbols that have not yet been made to refer to a specific entity. These symbols may be combined to form more complex symbol structures such as '**human(man, fred)**' or '**human(woman, X)**'. Supporters of the PSS hypothesis describe a PSS as being a system that produces through time an evolving collection of symbol structures and state that general intelligent systems, such as humans and any intelligent machine, must be physical symbol systems.

There are, however, many problems with symbolic artificial intelligence, both in terms of its ability to model human cognitive abilities and its ability to perform certain kinds of computation useful in practical situations. These problems include that of learning. It is simple for symbolic systems to learn by acquiring information already encoded in some propositional language, i.e. learn by being told. For example, in a Prolog expert system additional facts or rules can be directly coded into the database. However, symbolic models are very poor at learning on the basis of experience. Although much attention has been given to learning algorithms for symbolic systems such as ID3 [247], each of these generally operates in a specialized domain. In addition there is evidence that these algorithms are not as powerful as neural network learning algorithms [89], but this may depend on the domain that the learning algorithm is applied to. The lack of a general learning algorithm means that many symbolic models are hand-coded with a particular algorithm and problem domain in mind. In addition there seem to be many cognitive processes that do involve symbolic manipulation, in the sense of mapping one symbolic structure '**a**' into another '**b**', but are resistant to analysis at the level of the symbol. In Boolean terms this implies that there is no explicit specification of what the relation of '**a**' to '**b**' must be. A good example of such a process is face recognition, which might be thought of as a mapping that takes a set of symbolic structures representing facial features and produces a symbol representing the name of a particular individual. Humans can recognize the same face in a variety of contexts, from different angles, under different illuminations and with parts obscured by shadow or other objects. No known Boolean combination of features determines this mapping, yet humans can reliably recognize faces. This process is sometimes described as 'holistic' (a much abused word) as it resists the explicit decomposition of objects into component features and analysis at the level of the symbol.

Symbolic models tend to fail (brittleness) when presented with unusual, faulty or noisy input. Unless all the possible inputs the system can deal with are explicitly coded at the time of construction of the system, the system will be brittle. Many real-world tasks that engineers would like machines to be able to perform, and many human cognitive abilities, resist a complete specification of the inputs that a system is likely to meet.

These problems (among others) led many people to abandon the symbolic level description and take a different approach to modelling intelligent systems, the approach offered by connectionism.

1.2.2   The connectionist school

Researchers from this school believe that the level of the symbol is too high a level with which to design intelligent systems. They believe that instead of designing systems that perform computations on such symbols, we must instead design systems that perform computations at a lower level. This belief is well formulated in Smolenky's 'proper treatment of connectionism' [290]. In this article the 'subsymbolic hypothesis' states that a complete description of cognitive processing at the level of the atomic symbol does not exist and to give a full account of intelligent processes we must use a description that lies below the conceptual level.

1.3   Connectionist representations

One question in connectionist research is whether individual items in a network should be represented by a single unit (a localist representation) or whether the representation of an item should be distributed over a number of units (a distributed representation).

1.3.1   Localist representations

In these networks (often known as structured or spreading activation networks) each entity is represented by the activity of a single computing element. Every different item to be represented is mapped onto its own distinctive point or points in the representational scheme. These models tend to represent knowledge in structures similar to the semantic networks of classical AI in which concepts are represented by individual neurons and relations between concepts are encoded by weighted connections between those neurons. The activation level on each neuron generally represents the amount of evidence available for its concept in a given context. In these models the representations used by the system coincide with the computational tokens used by the system, so they are better regarded as being symbolic than subsymbolic. As each neuron represents a single semantically interpretable symbol it can be argued that there is nothing in these models that does not appear in classical symbolic processing models and localist systems are often described as being mere implementations of these.

Localist models tend to lack many of the properties possessed by distributed connectionist systems such as tolerance to noise and the ability to generalize. However, these networks are more readily understandable than networks using distributed representations, it is much easier to extract meaning from the activation pattern of the units as each unit represents a single concept. Localist models have been used to perform much interesting computational modelling, including the modelling of high-level inference [285].

1.4   Distributed representations

In these representations each thing to be represented by the system is represented by a pattern of activity distributed over many computing elements and each computing element is used in representing many different things [141]. However, there are many definitions of the notion of distribution in connectionist systems, some of which are listed below.

1.4.1   Microfeatures

Microfeatures are aspects of the domain that are much 'finer grain' than the top level items of the domain that are to be represented.   For example, to represent a type of '**room**' we can generate microfeatures that are usually found in various rooms such as '**sofa**', '**television**', '**ceiling**' and '**stove**'. Each different kind of room can then be represented by a distinctive pattern over these units. The pattern for '**lounge**' would be different from the pattern for '**kitchen**', but would still share some of the same microfeatures. Although microfeature based methods are one way of producing a distributed representation, they are not the only way. It is quite possible to develop a form of distributed representation where every item is represented by means of a semantically significant pattern over a set of units without individual units having any microfeatural significance at all [326].

1.4.2   Coarse coding

Another way to generate a distributed representation is by coarse coding [261]. This coding scheme places two conditions on the features assigned to individual units. Firstly, individual units should be given relatively wide receptive fields (the receptive field of a unit is defined as those features of the domain with which the activity of that unit is correlated). Secondly, the assignments should be overlapping; one unit may be involved in representing many different items. The representations of two different items can be superimposed on one another (as long as the number of units they both use remains relatively small compared with the number of units used to represent each item). This representation is robust, because of the overlapping nature of the receptive fields of the units no unit is crucial for a successful representation of an item. It also possesses semantically significant internal structure because the particular pattern used to represent an item is determined by the nature of that item, so similarities and differences among the items will be reflected by similarities and differences among the activation values of the units used in representing those items.

1.4.3   Extended and superposed representations

A more formal notion of distribution has been given by van Gelder [326], who described distributed representations with respect to their:

- *Extendedness.* For a representation to be extended the things being represented must be represented over many units, or more generally over some relatively extended proportion of the available resources in the system (such as the units in a neural network). Extendedness can produce a certain amount of robustness in a system. If an item is represented over several units in such a way that no particular unit is crucial to overall effectiveness of the system, then the system can be tolerant to some damage to the system's resources (such as the unit activations or weights in a network), or some noise in the inputs being processed by the system.

- *Superposition.* A representation can be said to be superpositional if it is representing many items using the same resources. Studies of the brain have shown many examples of this in human memory. For example, if the area of the brain used to recognize faces is partially damaged the ability to recognize all faces is damaged; there is no loss of the ability to recognize specific faces or classes of faces. This suggests that memories for faces are encoded in a superposed fashion in one localized region of the brain (containing millions of cells).

Representations in a standard feedforward network can be both extended and superposed. The representations can be thought of as consisting of two different parts [281]:

- A representation of the inputs coded in the weight representations, which stabilizes when the network has finished training. The representation of a particular input is then static and consists of the weights fanning out from it to each of the hidden units.
- A superposition of the inputs to the network, which results from summing the weights to each hidden unit multiplied by the activation of their relevant inputs and then applying the sigmoidal transfer function. This function is non-linear, so the effect of a particular input on the activation value of a particular hidden layer unit depends on the effects of the other inputs on this hidden layer unit.

The representation generated on a hidden layer unit in a feedforward architecture can be said to be extended as the representation of each input may be distributed over many hidden units and superpositional as each hidden unit can be representing several inputs. A method for measuring how extended and superpositional a particular representation is (and hence how distributed that representation is) has been developed [53].

Distributed representations have many advantages over symbolic or localist connectionist representations, including more efficient use of memory resources [192], the aforementioned generalization capabilities and the ability to model associative memory. In addition distributed representations give a new representational scheme with which to model computational processes. Classical symbolists (or the localist connectionists) believe that the mechanisms responsible for higher computational tasks (such as language processing) are supposed to divide into those responsible for syntactic (computational) tasks (such as parsing) and those which carry out semantically (representationally) oriented tasks (such as disambiguation). However, such a distinction is not present in systems using distributed representations. Often all the knowledge required to carry out the appropriate processing is contained in one set of connection weights. By ignoring this division between syntax and semantics distributed representations offer a new and different representational scheme that gives us systems radically different from conventional symbolic systems. However, some fundamental questions about distributed representations remain unanswered. These include:

- How distributed are the representations formed by connectionist networks? There is evidence that networks which have been usually assumed to form distributed representations, such as RAAMs [19], may not form very distributed representations at all [30].
- What are the limits of distribution in distributed representations? What are the upper and lower limits, and where do the representations formed by neural network training algorithms lie in relation to these limits?
- How does the level of distribution present in a distributed representation affect its fundamental computational properties?

1.5 Dynamic versus static representations

Classical symbolic systems perform their computations by passing through a sequence of states. Dynamic systems, however, are concerned with not only what states the system passes through but also how those states relate to time. They have many differences to standard computational systems, but perhaps the most important differences lie in their continuity in time and state:

- Time in a classical symbol system just refers to the ordering of the states that the system passes through. We cannot ask a question such as 'what computational state was the system in at time 2.5?' as each illusory 'time' step in fact just refers to an ordering of discrete states. With dynamical systems the equations that govern the behaviour of the system depend on time and the state of the system at any point in time can be calculated.
- Classical symbol systems change from one discrete state to another, there is no intermediate state lying between two successive states (it is not defined). On the other hand dynamical systems change in representational spaces that are continuous and hence any intermediate state can be represented.

The representations formed in many standard neural network models can be considered static. Although in one sense neural networks can be considered dynamical systems in that they are governed by continuous mathematical equations, Port and van Gelder [243] point out that many neural network researchers have not been utilizing concepts from dynamics:

> In standard feedforward backpropagation networks, for example, processing is seen as the sequential transformation, from one layer to the next, of static representations. Such networks are little more than sophisticated devices for mapping static inputs into static outputs. No dynamics or temporal considerations are deployed in understanding the behaviour of the network or the nature of the cognitive task itself ([243], p 32).

However, many neural network researchers are becoming interested in dynamics. For example many recurrent networks are being studied using techniques from dynamical systems theory (for an example of this applied to cellular neural networks see Chapter 5 of [1]).

1.6 Spiking neurons

In recent years there has been an accumulation of evidence that biological neurons use the timing of action potentials ('spikes') to encode information [4, 103]. This implies that in constructing artificial neural networks, analogue variables should be encoded by using the time differences between neuronal pulses, rather than by using the firing rate interpretation used by conventional neural network architectures. An explanation of this coding technique applied to VLSI neural networks is given in Chapter 4 of [1]. There is evidence [182] that a single spiking neuron can compute a function that a conventionally rate coded neural network needs hundreds of hidden units to encode. It remains to be seen whether the enhanced computational properties of networks using this coding technique force their general adoption by neural network users.

1.7 Challenges posed to connectionism by symbolic AI

Criticisms of the inability of connectionist systems to model certain capabilities of symbolic systems are outlined by Fodor and Pylyshyn [106] and Fodor and McLaughlin [105]. They advocate a 'language of thought' hypothesis, according to which cognitive activities require a language-like representational medium, which requires symbolic representations possessing constituent structure. They argue that connectionist systems cannot display capabilities such as the systematicity and compositionality displayed by symbolic systems (for a fuller explanation of these terms see Chapter 2, this volume) and are therefore incapable of performing complex structure-sensitive tasks. The following problems also face connectionist modellers if they are ever to argue that their models possess the power of symbolic computational systems.

1.7.1 Structural systematicity

A number of different definitions of structural systematicity have been given. Three levels of systematicity from weak to strong were defined by [126], but perhaps the most precise definition of six levels of systematicity has been given by Niklasson and van Gelder [213]. These levels are:

- Where no novelty is present in the test set presented to the network. Every test term presented to the network has already appeared in the set of input patterns that the network has been trained on.
- Where novel input patterns appear in the test set, but all the individual inputs in these patterns have at some time appeared in the same positions in the training set.
- Where the test patterns contains at least one pattern that has an input in a position in which it never appeared in the training set. For example, a network could be trained on a selection of input patterns in which a particular input was never present in the first input position of the patterns used for training and then tested on a set of patterns where that input was present.
- Where novel inputs appear in the test set. The test set contains at least one input that did not appear anywhere in the training set.
- Where there is novel complexity in the test set. To display this form of complexity a network would have to be capable of being trained on patterns with n inputs in the training set and then correctly process patterns with $n + 1$ inputs in the test set.
- Where the test set contains novel inputs and novel complexity. Both the third and fourth points above would have to appear in the test set.

Symbolic systems can display all these levels of systematicity as they are not dependent on the makeup of a training set to form their representations. Hence they can generate and process new items whilst only being restricted by physical constraints, such as the size of a computer's memory. Work by connectionists has shown that systematicity up to the second level can be performed on distributed representations without recourse to the symbolic level. These tasks include language translation [41, 67], unification [56], resolution [54], performing De Morgan's laws [215], parsing [95], inferring verb meaning [26] and representing the spatial orientation of objects [40]. Recent connectionist research indicates that this could be extended to the third level described above. In performing simple logical operations on distributed representations [213] systematic performance of a neural network both when exposed to novel inputs and when exposed to inputs appearing in novel positions has been demonstrated. This work, and other work using tensor-based representations, has pushed back the frontiers of the representation of systematic structure in connectionist systems. Recently Robert Hadley, a fierce critic of the work described above [124], has produced a model that, he argues, displays a stronger form of systematicity than that displayed in [213]. In this model [125], a network using Hebbian and Kohonen training can generalize to sentence constituents not encountered in training and can also generalize to sentences with deeper levels of embedding than those encountered in training. However, since this model also has a layer that makes use of symbolic message passing, it is unclear whether it truly refutes the systematicity criticisms levelled at connectionist systems.

1.7.2   Macro-systematicity

One form of macro-systematicity is concerned with representations being portable within the same computational system. If we consider a neural network that represents the two concepts '**Fred loves Mary**' and '**Mary loves Fred**', it would be displaying macro-systematicity if in both cases the '**Fred**' component had the same (or similar) distributed representation. This has been demonstrated by [282], where the distributed representation of a particular item formed on the hidden layer of a feedforward network could be extracted as a vector ξ and used successfully in another context.

This form of macro-systematicity is displayed in symbolic systems, as symbols with the same name have the same representation and are readily interchangeable in these systems. It could be argued that because of this, symbolic systems display a 'stronger' form of macro-systematicity. However, it is possible that this is trying to force connectionist systems into a symbolic straitjacket and that as long as distributed representations can be extracted from one context and successfully used in another, then this form of macro-systematicity has been displayed by connectionist systems.

In another form of macro-systematicity representations would be portable between different computational systems. In a connectionist context, this would involve being able to take the vectorial representation of an entity formed on the hidden layer of one network and use it (still preserving its meaning) in another network. The distributed representations formed in connectionist networks depend on the training patterns applied and the initial states of the system (such as the set of random weights used before training commences) and hence are not re-usable when transferred to another network trained under different circumstances. This limits the portability of connectionist representations as they only have meaning within the system that generated them. This criticism may also be true of symbolic systems. In these systems symbols obtain their meaning from their relationship to the other symbols within that system. If a symbol is extracted from one symbolic computational system and then an attempt is made to use it in another different computational system it may not convey the same meaning.

1.7.3   Variable binding

Many leading researchers have argued that the ability to handle variables is essential for connectionist systems if they are ever to perform complex reasoning tasks. There are several main points to this argument:

- Variable binding is essential in creating dynamic structures in connectionist networks [102]. For example, if we have a system that can reason with data consisting of '**white cats**' and '**black dogs**', we must have a way of associating (binding) '**white**' with '**cat**' and '**black**' with '**dog**' (such as '**colour(X, white), animal(X, cat)**').
- Variable binding is essential in modelling the human language acquisition faculty [233] as it imposes constraints on rule matching needed when modelling certain aspects of past-tense learning.
- Variable binding mechanisms are needed to achieve greater computational power in connectionist systems [296]. Early connectionist models avoid variable binding by reasoning with one object at a time or using multiple instantiations of the same object [285]. This approach limits the expressive power of these systems. Instead of having a general rule such as '**mammal(X) :- gives milk(X)**', i.e. '**X is a mammal if X gives milk**' we must have many instances of more specific rules, such as '**mammal(dog) :- gives milk(dog)**', '**mammal(cat) :- gives milk(cat)**', '**mammal(horse) :- gives milk(horse)**' etc.

Many of the implementations of variable binding using connectionist representations have been part of larger inference systems such as production systems. Most have been localist models [27, 10, 199, 36, 314, 294, 186] and therefore subject to the criticisms usually levelled at such models. A limited form of variable binding was implemented in a system using a mixture of localist and distributed representations, the Distributed Connectionist Production System [316]. However, in this system variable binding could only be carried out on the variable in the first argument position in the rules represented. In addition, although the representations in the system were distributed, rules application was performed in a serial symbolic fashion, relying on the settling of a Boltzmann machine to extract representations on to a symbolic level. For a description of a connectionist network that performs variable binding and inference using a mixture of localist and distributed representations see Chapter 5 of this volume. The first system to perform variable binding on fully distributed representations was a system performing a complex form of variable binding called unification [55, 56].

1.7.4   Performance of inference

Although there have been many attempts to produce connectionist systems that perform the kinds of inference processes performed by symbolic systems

[170, 144, 285], these models used localist connectionist representations and so are subject to the criticisms attached to this type of representation discussed above. Resolution [259] is a powerful formal inference rule which can find new facts and relationships that follow logically from existing facts and relationships. It is a derivation of the logical rule '**from a implies b and b implies c, deduce that a implies c**'. Applications of resolution-based inference systems include the reasoning performed by various expert systems and the design and validation of logic circuits [154]. Resolution has been performed directly on distributed representations [54], but this model can only process a fixed set of variables (unlike symbolic systems which can create them dynamically). In addition, in a resolution proof often one of a series of resolution steps will fail, forcing the system to retrace its steps and then attempt a different series of resolutions. This entails the provision of a backtracking mechanism (as is present in Prolog interpreters), which is not present at the moment in the model described in [54], but could possibly be provided at some time in the future using some form of RAAM [19] to provide the stacking facility required.

1.7.5   Limits on the productivity and width representations

The productivity of a system refers to the ability of that system to generate and correctly process items from an infinite set. Because of the possibility of nesting of arguments, structures can be produced the size of which cannot be pre-determined. The property of productivity is displayed by symbolic systems which can produce and process an (almost) infinite set of propositions from a small set of symbols using recursive structures (only being limited by physical constraints such as memory size). Neural networks have a limited set of inputs and can only cope with a pre-specified level of recursion. Because of this, the set of input patterns that can be correctly generated and processed is completely specified when the network is constructed. Recursive connectionist architectures such as RAAMs [19] and XRAAMs [172] do exist, but the structures represented within them cannot be modified 'on the fly' to produce new structures when required. However, there is also evidence that humans exhibit a limit on the depth of recursion they can process. Considering the sentence 'John told Mary that Betty told Fred that Sally told Frank that Jim went home', Dyer [94] points out that:

> Most people, upon hearing such a sentence out loud, protest that they cannot keep straight who is telling whom and immediately recall it as 'Several people telling other people that Jim went home.' ([94], p 399)

The number of elements in a representation affects the level of relation that the representation can represent. The number of components can be thought of as the number of facets of a situation that can be viewed simultaneously. There is considerable evidence [127] that most humans can reason using representations with at most five components, such as a predicate (the first component) with four arguments (the four other components). An example of this would be a predicate '**proportion(A, B, C, D)**', which expresses a relation between four variables such that **A/B = C/D**. In this predicate it is possible for a human to predict how any variable will change in relation to one or more of the others. The psychological existence of the processing of rank six representations (such as a predicate with five arguments) by humans is speculative and if it exists it probably does so only for a minority of adults. This implies that insofar as human cognitive modelling is the field of interest, a network would only have to be able to process inputs representing relations with a maximum arity of five.

The seriousness of these deficiencies depends very much on the perspective that neural network systems are viewed from. From the perspective of a computer scientist attempting to implement all the capabilities of a symbolic system using a neural network, these restrictions are serious. From the alternative perspective of a researcher attempting to model human cognitive abilities, they may not be such serious criticisms.

1.7.6   Performing dynamic inferencing

Symbolic models have the ability to perform dynamic inferencing from an initial set of bindings by applying their rules to infer novel intermediate states having new bindings. Further inferences can then follow repeatedly from the new intermediate states until the desired state is reached. This ability is crucial when a system cannot reach a result in a single step. Marker-passing networks and recent localist networks can hold variable bindings and propagate them in turn for inferencing. In this way they can also perform dynamic inferencing, though their inference rules are generally limited in complexity relative to symbolic models. Traditional pattern-transformation distributed networks cannot perform dynamic inferencing, because they transform the input (or set of inputs) to the output in a single step.

1.7.7   Parallelism at the knowledge level

Marker-passing and localist networks are able to explore multiple solutions in parallel because alternative interpretations are represented by markers or activation patterns across different local areas of the network. This is crucial because with dynamic inferencing there are often a very large number of alternative solution paths, especially in language understanding and planning. In contrast, although distributed networks update their units in parallel, they are serial at the knowledge level because they represent all dynamic knowledge in a single set of units on the output or in a hidden layer and so cannot make dynamic inferences from more than one potential solution at a time. Conventional feedforward networks using distributed representations can sometimes hold ambiguous solutions in a single blended activation pattern on their hidden units, but it remains to be seen how far such patterns can be extended to simultaneously hold and make dynamic inferences from multiple, never-before-encountered solution paths.

1.7.8   Performing meta-reasoning

Symbolic models can reason and operate on knowledge from many different substructures of their program, so long as that knowledge is represented by globally interpretable symbolic structures. Localist connectionist and marker-passing networks can do this if they employ a separate high-level symbolic program to interpret and work with the results. Pure distributed networks and localist networks attempt to complete their tasks entirely within the network and so cannot perform meta-reasoning by resorting to symbolic code. This 'monolithic' nature prevents these forms of networks re-using 'chunks' of knowledge in different places, in contrast to the ease with which this can be done in symbolic systems such as expert systems.

1.8   Other criticisms of connectionism

There are other criticisms levelled at connectionism, from both neuroscientists and engineers trying to solve practical problems using neural network systems.

1.8.1   Biological plausibility

Although neural networks may initially seem biologically plausible as they 'take inspiration from the brain' many of the established techniques used by neural network modellers (especially the backpropagation of errors) are not biologically realistic. Some of these deficiencies are noted by Crick [76], who states that 'in the real brain the outputs of single neurons are either inhibitory or excitatory, but not both as is seen in many neural network models'. If the brain used an algorithm in some way similar to the backpropagation of errors techniques used in artificial neural networks, this would require transmission of information back from the synapses and along the axon. This is highly unlikely to occur in real brains. Some researchers argue that backpropagation is only one of many different procedures that can perform gradient descent and is merely a useful programming technique that is not intended to model any aspect of human neurophysiology [142]. Other researchers attempt to produce neural-network-based cognitive models that use more biologically plausible techniques [235] or argue about the actual training methodology used in connectionist cognitive modelling [157].

Related to the above criticism is the notion that backpropagation and related algorithms require the sharing of information among neurons. Each neuron needs to know the error signals attached to neurons in another layer of the network in order to modify its weights, i.e. the weight update term between the hidden and output layers requires information from both output unit k and hidden unit j, and between the input and hidden layers requires information from input unit i and hidden unit j. However, a strictly local version of backpropagation has been developed [100] which performs local calculations only. In this variant of backpropagation the network has three types of unit:

- Cortical units, which sum their inputs and send the result to the units above them, which are single synaptic units.
- Synaptic units, which receive an input signal from a single cortical unit, apply an activation function to that signal and then multiply the result by a weight. This result is then sent on to a thalamic unit.
- Thalamic units, which compare the computed output with the target value and generate an error signal for the appropriate thalamic unit lying below them.

In artificial neural networks the computer calculates the error between the network's actual output and its desired (target) output. In the brain 'teacher' neurons would be required to calculate this error: no evidence has been found of these yet.

As each output synaptic unit receives the error signal from the single thalamic unit above it, multiplies this signal by its weight, multiplies this again by the derivative of its activation function, and then sends it to the output cortical unit below, this can be considered a strictly local form of backpropagation. For a more complete description of this algorithm see [101]. However, there is some evidence of the biological plausibility of neural network models as some studies have found a correspondence between the neurophysiological results relating to the receptive fields of real neurons and the hidden unit feature detectors developed by artificial neural network models of these same neurons. For a more in-depth discussion of the biological plausibility of neural network learning algorithms see Chapter 11 of this volume.

Another criticism of biological plausibility comes from our knowledge of evolutionary development. Would organisms develop one form of learning, that displayed by the immune system, and then in some way fail to integrate this in some way in another form of learning, that which we currently believe takes place at the synapse? Single-celled animals that possess no neurons or synapses can still learn, albeit only simple tasks such as moving toward or away from a light source to seek food or avoid a noxious substance. Where does our learning theory based solely on the modification of synaptic weights lie in relation to these observations?

1.8.2 Network scalability and training times

Many neural network techniques were initially developed on small problems which were small enough to understand but difficult enough that their solution significantly enhanced understanding of connectionism. Many of these techniques have been successfully scaled up to deal with problems encountered by commerce and industry. However, not every problem encountered in dealing with small systems can be scaled up easily. In the words of Minsky and Papert [197]:

> Looking at the relative thickness of the legs of an ant and an elephant reminds us that physical structures do not always scale linearly: an ant magnified a thousand times would collapse under its own weight. Much of the theory of computational complexity is concerned with questions of scale. If it takes 100 steps to solve a certain kind of equation with four terms, how many steps will it take to solve the same equation with eight terms? Only 200, if the problem scales linearly. But for other problems, it will take not twice 100 but 100 squared. ([197], p 262)

Training times for neural networks attempting to solve large problems can be extensive, running into the order of hours or days for large models. Although fast training algorithms exist, such as the scaled conjugate gradient algorithm for feedforward networks (described in Chapter 6 of [1]), all network training algorithms do not scale well on serial hardware. The use of parallel hardware could partially remedy this situation. Some speed limitations could in theory be overcome by implementing neural networks directly in silicon. There are, however, some difficulties to be overcome, most notable being the limitations in accuracy of weight and transfer function implementation in silicon [24]. In addition there are problems connecting neurons together for large collections of neurons, as the actual physical act of creating physical connections becomes hampered by space and heat dissipation requirements. Novel technologies such as optical implementation of the network weight matrices may supply an answer to the connectivity problems, but as yet these technologies are still at an early stage of development.

1.9   Conclusion

Members of the symbolic AI camp criticize the attempts by connectionists to model the abilities of symbolic systems. Connectionists on their part are hard at work attempting to build systems that attempt to answer these criticisms. Networks using dynamic representational schemes and/or the added computational power of models based on spiking neurons may in future demonstrate abilities that nullify the criticisms levelled at connectionist models. However, it could be argued that attempting to model symbolic systems is a rather arbitrary target set for the connectionist community by the symbolic camp, which should be ignored and bypassed in the search for more powerful connectionist architectures. It is also argued that the limitations of both types of system should be avoided by producing hybrid systems which give the advantages of each type of model and avoid the disadvantages (for an overview of several hybrid systems see [300]). It is perhaps prudent to realize the limitations of both the symbolic and the connectionist camps and consider that in the not too distant future both paradigms may have to be abandoned in favour of a model neither side can yet envisage.

Chapter 2

Representing Structure and Structured Representations in Connectionist Networks

Lars Niklasson[1] and Mikael Bodén[2]
University of Skövde, Sweden
[1] lars.niklasson@ida.his.se
[2] mikael.boden@ida.his.se

2.1 Introduction

Connectionist networks have earned recognition in many domains that can be characterized as hard or impossible to explicitly formalize, e.g., driving cars [242], emotion recognition [74] and pronunciation [339]. Connectionists such as Rumelhart and McClelland [193] have also claimed that their networks can exhibit behaviors that can be described by a set of formal rules, without actually implementing explicit rule following. The radical implication of this claim is that connectionism does not appear to neatly line up with the classical view of the cognitive architecture, i.e., the computational, the representational and the implementational levels (described by Marr [188], Pylyshyn [246], Newell [211], Anderson [12]).

The intention of this chapter is to investigate a number of different aspects of this claim. We will compare two computationally equivalent systems. The behavior of both these systems can be described by a set of general principles, i.e., rules, but the representations and processes used by these two systems are fundamentally different. The two systems we will contrast and debate will be a physical symbol system [210], hereafter referred to as a PSS, and a connectionist system based on the use of non-symbolic representations and processes sensitive to spatial structure, hereafter referred to as a non-symbolic system (NSS).

The task selected for the two systems is to classify simple tree-structures (figure 2.1) according to certain criteria, such as the balance or depth of the tree-structure.

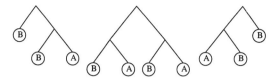

Figure 2.1. The represented world.

We could describe this task by formulating a set of formal rules or functions. An example function is shown below, which returns 'left' if the tree-structure is left balanced, 'right' if it is right balanced and 'well' if it is equally balanced

function balance (X) returns {left, right, well}.

If the two systems return the same classification for all the tree-structures, the systems are said to be computationally equivalent. This is, however, only a very weak form of equivalence, since we are not taking into account *how* the functions generate the output depending on the input. We are also not taking into account the aspect of time; one of the systems might return a value in seconds, and the other one in years. With this we want to highlight that arguments such as 'anything your system can do, ours can too, so what's new with your system?', may not be valid arguments when comparing the two systems we are interested in (this point is also made by Rumelhart and McClelland [262], in their response to Broadbent [47]).

2.2 Representations

The computational level of the classical cognitive architecture describes *what a system computes*, and *why*. At this level, the two systems of interest here compute the same function, i.e., their external behavior is indistinguishable. We have to move one level down from the computational level to detect the differences between the two systems. At this lower level, the representational level, the interest is focused on *what the system is actually doing* in order to compute the function it is designed for.

The question follows; what is a representation? In an extensive survey Palmer [225] argues that a representation is something that stands for (or represents) something else. This means that we have to define two different worlds; the represented world and the representing world. Palmer points out that a 'representation' really is a representational system including specifications of:

(i)   What the represented world is.
(ii) What the representing world is.
(iii) What aspects of the represented world are being modeled.
(iv) What aspects of the representing world are doing the modeling.
(v)   What the correspondences between the two worlds are.

The represented world is here a domain containing two atomic tokens (*A* and *B*). This world is the same for both the systems we envision. The aspect of the represented world that we want to model is that these two tokens can appear in structured combinations, in accordance with a simple grammar. The differences between the two systems are due to the choice of representing worlds.

The first criterion that the two systems need to meet is that their representing worlds can model (or represent) the represented world, here a set of tree-structures.

2.2.1   Representing structure in a PSS

A PSS uses a representing world containing 'symbols', e.g., *a* and *b* (see point (ii) above). These symbols can be realized in a lower-level physical implementation, such as metric vectors. However, since the lower level is an implementation of the higher level (in the sense that the lower level cannot contribute in anyway to the understanding of *what the system is doing*), it can be discarded.

The aspect of the particular PSS we are interested in here that is doing the modeling is that the symbols can be organized according to a grammar equivalent to the grammar for combining the external tokens (see point (iv) above), and the correspondence between the two worlds is the mappings $A \rightarrow a$ and $B \rightarrow b$ (see point (v) above and figure 2.2).

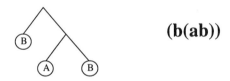

Figure 2.2. The represented world (left) and the representing world (right).

Since the representing world here is governed by an equivalent grammar to the represented world, it has the capability to represent the external tokens in isolation or combined into a structure. In fact, the definition of a PSS hinges on the ability to generate representations (i.e., organize the representing world) for the complex expressions of the represented world in this concatenative way. Fodor and Pylyshyn [106] mention that this is one of two main pillars of the classical view of the cognitive architecture, stating that a PSS must have:

> (1) Combinatorial syntax and semantics for mental representations ... in which (a) there is a distinction between structurally atomic and structurally molecular representations; (b) structurally molecular representations have syntactic constituents that themselves are either structurally molecular or structurally atomic; and (c) the semantic content of a (molecular) representation is a function of the semantic

contents of its syntactic parts, together with its constituent structure. ([106], pp 12–13)

In our example, it is the case that there is a distinct difference between, *a* and $(b(ab))$. We can also see that the molecular representation $(b(ab))$ has both complex and atomic building blocks, or constituents. The semantic content of $(b(ab))$, i.e., what it refers to in the represented world, depends on what *a* and *b* refer to combined with how they are organized in this particular configuration, compared to $((ba)b)$.

In a classical machine, these features of the representing world are essential for explaining mental processes. We can explain how this type of machine works by referring to the second pillar of the classical view of the cognitive architecture:

> (2) Structure sensitivity of processes.... Because Classical mental representations have a combinatorial structure, it is possible for Classical mental operations to apply to them by reference to their *form*. ([106], p 13, emphasis supplied)

This means that classical mental representations must have some sort of internal structure in order to allow mental processes, the question is which *kind* of structure?

2.2.2 Structured representations in a PSS

What does it mean to have 'structured representations'? Here we mean that the representing world has some properties that can be explicitly formalized by using rules or algorithms. This relates to Palmer's point (iv) above; the aspects of the representing world that are doing the actual modeling are governed by a set of formal principles.

A PSS is firmly based on the notion of *concatenative compositionality*, i.e., a combinatorial syntax resulting in representations which have a constituent structure, which is defined as follows:

> [F]or a pair of expression types E1 and E2, the first is a classical constituent of the second only if the first is tokened whenever the second is tokened. For example, the English word 'John' is a classical constituent of the English sentence 'John loves the girl' and every tokening of the latter implies a tokening of the former (specifically, every token of the latter contains a token of the former; you can't say 'John loves the girl' without saying 'John'). ([105], p 186)

If related expressions are constructed by using the same constituents, according to Fodor and Pylyshyn, this could explain why humans possess a systematicity of thought.

> What does it mean to say that thought is systematic? Well, just as you don't find people who can understand the sentence 'John loves the girl'

but not the sentence 'the girl loves John', so too you don't find people
who can *think the thought* that John loves the girl but can't think the
thought that the girl loves John. ([106], p 39, original emphasis)

Both 'John' and 'the girl' can move around in the thoughts and accept
different roles. The PSS has a structured way to represent related thoughts, i.e.,
a *systematicity of representation*.

2.2.3   Structure sensitivity in a PSS

Constituently structured representations are, as mentioned earlier, also important
for explaining classical mental processes. If the syntax of the representational
level can be made to parallel the semantics of the computational level, it is
possible to define a syntactically driven machine which operates on the form of
representations and not on their content.

If, in principle, syntactic relations can be made to parallel semantic
relations, and if, in principle, you can have a mechanism whose
operations on formulas are sensitive to their syntax, then it may
be possible to construct a syntactically driven machine whose state
transitions satisfy semantical criteria of coherence. ([106], p 30)

This means that *what a PSS is doing is symbol manipulation*. The input to
a system of this kind is a constituently structured representing world, which is
manipulated by structure sensitive processes in order to return the output from
the system. It is left to the reader to use a programming language and construct
a system that accepts constituently structured symbol strings such as $(b(ab))$
and returns their classification such as 'right balance'.

The challenge posed by Fodor and Pylyshyn to connectionists was to explain
structure sensitivity, without resorting to constituently structured representations.
If connectionists failed to respond to this challenge, their attempts to provide
an alternative to the PSS would be futile since humans necessarily exhibit
a 'systematicity of thought'.   If they responded, but responded by using
constituently structured representations, then their alternative would be nothing
else than an implementation of a PSS, Fodor and Pylyshyn argued.

Let us now turn to how connectionists have attempted to respond to this
challenge, by presenting different kinds of connectionist representing worlds.

2.2.4   Representing structure in connectionist networks

Connectionists have used a whole arsenal of different types of representing
world, all based on algorithms for calculating the state of activation of simple
processing units. The interpretation of these units stretches from localist, via
microfeatural, to non-symbolic representations. We will here briefly debate the
merits of each of these with regard to the issue at hand.

2.2.4.1 Non-compositional connectionist representations

Fodor and Pylyshyn discussed the non-compositional type of connectionist representation and warned that it is easy to overlook the difference between arrays of symbols and the use of labels in connectionist networks.

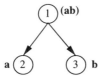

Figure 2.3. A localist representation.

This can be exemplified with a simple network of interconnected units (see figure 2.3). One should not, Fodor and Pylyshyn pointed out, be misled to think that this graphlike figure could be interpreted such that *a* is a constituent of *ab*. It can only be interpreted as unit 2 is causally affected by unit 1. The symbolic representation literally contains the constituents as tokens, whereas this type of connectionist representation is made up of causal connections between *atomic* units. This means that only the symbolic representation allows structural relations between its constituents (i.e., there is no structure in the atomic unit representing (*ab*)), which we have seen is the explanation of the behavior of the classical machine. The main problem (from a compositional point of view) with this type of representing world is that all the units are atomic and cannot be combined into representations for complex expressions. So if we want to model the aspects of the represented world that some tokens can be organized in more or less complex structures, that is harder to accomplish in this representing world. If, for instance, we want to represent not only (*ab*) but also (*ba*), we need to introduce the new atomic binding unit (*ba*) and connect it to the units labeled *a* and *b*. It appears that this localist representational scheme does not allow structural relations between constituents. The potential problem (highlighted by Fodor and Pylyshyn) is that the *labels* on the units (which appear to have a constituent structure) do not play any part in the behavior of the connectionist machine. This means that it appears to be expecting a lot of our high-level functions to classify networks of atomic units only connected by causal links in a novel way (compare the two three-unit networks needed to represent (*ab*) and (*ba*)).

Assuming that we could build a representational theory that could use these atomic representations (without the notion of constituents and the classical mode of composition) the question is whether that theory could account for structure sensitive processes in a novel way. So from the current perspective, this representing world does not seem to have the means for representing or having the structure necessary to generate behavior indistinguishable from a PSS.

2.2.4.2 *Weakly compositional connectionist representations*

In the localist representational scheme one concept was represented by one unit, in a one-to-one mapping. It could also be the case that the concept is represented in a distributed fashion, i.e., using a number of units, in a one-to-many mapping. This form of distributed representation is still local in the sense that it is possible to localize the concept to its uniquely allocated units.

When connectionists talk about distributed representations, they normally have a many-to-many mapping in mind, i.e., that one concept is represented by many units and each unit takes part in the representation of many concepts. Two basic forms of distributed representation can be identified; microfeatured and non-symbolic representations. When using microfeatures each unit (or set of units) is seen as referring to one property of the concept. This kind of interpretation is not possible with non-symbolic representations. Instead of concentrating on individual units, the interest here is only what the whole set of units represents.

Fodor and Pylyshyn argued that microfeatured representations should not be seen as having constituent structure, for the following reasons. A sub-symbolic representation of a high-level concept (e.g., BACHELOR) is made up of a description of sub-conceptual features (e.g., [man, woman, human, animal, adult, youngster, married, unmarried]), typically in the form of a vector (e.g., BACHELOR $= [1, 0, 1, 0, 1, 0, 0, 1]$ or WIFE $= [0, 1, 1, 0, 1, 0, 1, 0]$).

The problem that Fodor and Pylyshyn identified for this type of representation was that it has no combinatorial structure. One should not mix up the point that a microfeature (e.g., +*Has-Handle*) is a part of the distributed representation of a high-level concept (e.g., CUP), with the question of combinatorial structure. The microfeature +*Has-Handle* is not part of (i.e., a constituent) of the expression CUP:

> any more than the English phrase 'is an unmarried man' is part of the
> English phrase 'is a bachelor'. ([106], p 22)

The lack of 'real' constituent structure means that it is impossible to have relations between constituents. This might lead to troublesome situations, e.g., if we want to represent the situations 'John loves Mary and Bob hates Sue'. A micro-featured vector, e.g., [+John, +loves, +Mary, +Bob, +hates, +Sue], does not transparently describe the situation. It is, for instance, here impossible to know who the subjects of the situations are. There is a big difference between being simultaneously active and in construction with each other ([106], p 25). This problem could be overcome by also encoding that information in microfeatures, e.g., [+John-subject, +Mary-object, +Bob-subject, +Sue-object]. Still we have problems to identify which subject is in which relation and with which object. Again, this is possible to solve by adding a feature that encodes that situation, e.g., [+John-subject-loves-Mary-object, +Bob-subject-hates-Mary-object].

The problem is that constituents like 'John' cannot easily move around in the representations and accept different *roles*. This role binding is instead achieved by introducing microfeatural binding units. Fodor and Pylyshyn suggest one alternative to this explosion of atomic features; a combinatorial syntax and semantics for the features, i.e., to become a classicist.

The introduction of these binding units has another effect on this representational scheme, it becomes context sensitive (highlighted by Smolensky [290], among others).

> In the symbolic paradigm, the context of a symbol is manifest around it and consists of other symbols; in the subsymbolic paradigm, the context of a symbol is manifest inside it and consists of subsymbols. ([290], p 17)

This means that this representational scheme could be departing from the demand of unique composition and decomposition of representations for complex expressions. If the representation of 'cup without coffee' is subtracted from 'cup with coffee', we could be left with a feature vector for 'coffee' but in the context of being inside a cup.

> These constituent sub-patterns representing coffee in varying contexts are activity vectors that are not identical, but possess a rich structure of commonalities and differences (a family resemblance, one might say). ([290], p 17)

Can the family resemblance be used for systematic processing, i.e., can this type of representation form the basis for systematic processing? Fodor and Pylyshyn stated that,

> [S]ince such proposals aren't generally elaborated, it's unclear how they're supposed to handle the salient facts about systematicity and inference. ([106], p 45)

One could be inclined to agree with Fodor and Pylyshyn here. It is hard to see how one could implement structure sensitive processes in any novel way when operating on microfeatured vectors. It is not entirely obvious that even a successful demonstration of structure sensitivity would avoid the implementation aspect of the challenge.[1]

Instead of relying on simple addition and subtraction of microfeatured vectors, some connectionists turned to the feature that is the basis for connectionist networks; a set of weights for associating vectors.

---

[1] See Pinker and Prince [233] for a variation of this view.

2.2.4.3   Strongly compositional connectionist representations

Could association be a connectionist variation of classical compositionality? Van Gelder [325] argued that concatenation is not the only possible means for generating representations for complex expressions, stating that:

> In the general case, all that is required of a mode of combination is that we have systematic methods for generating tokens of compound expressions given their constituents, and for decomposing them back into those constituents again. From the point of view of generating expressions satisfying the abstract constituency relations, there is no inherent necessity that those methods preserve tokens of constituents in the expressions themselves—rather, all that is important is that the expressions exhibit a kind of functional compositionality. ([325], p 359)

Since we have granted that distributed connectionist representations are context sensitive, the question is, what are to be considered *the* representations for a constituent. Do we have to use the stable symbolic constituents or is it possible to use the more approximate context-dependent connectionist representations? According to Fodor and McLaughlin [105], it could turn out that:

> complex superpositional vectors will have it in common with Classical complex symbols that they have a unique decoding into semantically significant parts. Of course, it could also turn out that they don't, and no ground for optimism on this point has so far been supplied. ([105], p 199)

This means that connectionists could solve the problem if they somehow could categorize representations that have a family resemblance into significant categories. This would allow the definition of stable classes (say, CUP) and preservation of contextual difference (COFFEE CUP, TEA CUP, AMERICA'S CUP, DAVIS CUP). Let us now investigate this opening for connectionist representations by presenting two compositional schemes for generating representations for complex expressions; tensor product representations and the recurrent auto-associative memory (RAAM).

2.2.4.4   The tensor product representational scheme

The basic idea behind the tensor product representational scheme (Dolan and Smolensky [90], Smolensky [292]) is that any concept with constituent structure (e.g., the sentence 'John loves Mary') can be represented by combining the representations for all the constituents, associated with the role the constituent plays in the complex concept (e.g., [John/agent] + [loves/relation] + [Mary/object]). This means that we will have a representational scheme

Role   = $\begin{bmatrix} 1 \\ 0 \\ 0 \end{bmatrix}$ $\begin{array}{ccc} 1 & 0 & 0 \\ 0 & 0 & 0 \\ 0 & 0 & 0 \end{array}$ = [1 0 0 0 0 0 0 0 0]

Agent

$\underline{1\ 0\ 0}$ = Filler
John

Role   = $\begin{bmatrix} 0 \\ 0 \\ 1 \end{bmatrix}$ $\begin{array}{ccc} 0 & 0 & 0 \\ 0 & 0 & 0 \\ 0 & 1 & 0 \end{array}$ = [0 0 0 0 0 0 0 1 0]

Object

$\underline{0\ 1\ 0}$ = Filler
Mary

Role   = $\begin{bmatrix} 0 \\ 1 \\ 0 \end{bmatrix}$ $\begin{array}{ccc} 0 & 0 & 0 \\ 0 & 0 & 1 \\ 0 & 0 & 0 \end{array}$ = [0 0 0 0 0 1 0 0 0]

Relation

$\underline{0\ 0\ 1}$ = Filler
loves

Tensor
John-loves-Mary $\begin{array}{ccc} 1 & 0 & 0 \\ 0 & 0 & 1 \\ 0 & 1 & 0 \end{array}$ = [1 0 0 0 0 1 0 1 0]

Figure 2.4. Role, filler and tensor representations of 'John loves Mary'.

with three types of representation; role representations, filler representations and role/filler representations (the tensor). For an example see figure 2.4.

It is also possible to represent complex situations like 'Bob knows [that] John loves Mary'. This can be achieved by combining the tensors for the two parts not by simple addition but by generating a new tensor, i.e., of rank three, where the role and filler vectors are combined with a vector representing whether the role/filler pair belongs to the first (i.e., 'Bob knows . . . ') or second (i.e., 'John loves Mary') part of the sentence. It deserves to be noted that this particular example could be seen as an implementation of the classical concatenative mode composition, since there are clear tokens (e.g., [1, 0, 0] which is interpreted as 'John', in the agent position of the John-loves-Mary tensor) present in the tensor; it has a constituent structure.

There are two ways a tensor can be decoded; the exact unbinding procedure and the self-addressing procedure. The self-addressing decoding procedure works for any kind of role vector, but will only result in an accurately decoded filler representation (apart from an overall magnitude factor) if the role vectors are all orthogonal, i.e., if the inner products between the vectors are all zero. If the role vectors are non-orthogonal the decoded representations will be affected by intrusion from other role vectors. If the role vectors are linearly independent, the exact unbinding process might be used instead. Instead of using the actual role vectors for the decoding process, unbinding vectors are used. One approach, suggested by Smolensky [292], to find the correct unbinding vectors is to train a separate network (e.g., by using the Widrow–Hoff, or delta rule[2]) to associate each role vector with one unit in a set of output units (one for each role). When the network has learned to correctly classify the role vectors, the weights are made bi-directional. When a certain role vector is supposed to be decoded, its local output unit is activated and the unbinding vector is generated over the input units.

This type of representation seems to have an advantage over the weakly compositional scheme, outlined earlier. It appears that it allows a constituent to move around and accept different roles. This is also noticed by Fodor and McLaughlin, who state:

> It's not, at least for present purposes, in doubt that tensor products can represent constituent structure. The relevant question is whether tensor product representations have constituent structure . . . ([105], p 200)

We have here chosen to differentiate between representing structure and structured representations, because our intention here is to argue that it is possible to represent a structured world in a *structured manner, without using constituent structure as a basis.* So, why have structured representations become synonymous with *constituently structured representations*?

> Largely because concatenative modes are so common and familiar, we tend to lose sight of the fact that concatenation, . . . , is not the only way of implementing the combination of tokens to obtain expressions. ([325], p 359)

The question here is, does the tensor have sufficient structure to allow structure sensitive processes (i.e., other than simple encoding/decoding)? One problem with applying structure sensitive processes on tensors is that the size of the tensor varies with the complexity of the represented expression.[3] This might be the reason why the attempts to show that tensors are susceptible to structure sensitive processes have been conducted on combinatorial domains containing the same complexity for all the expressions represented (e.g., Smolensky and Brousse [291], Phillips [230]). Another potential problem might be that a successful demonstration of structure sensitive processes applied to a tensor would be classified as simple constituently structured sensitivity. However, a representation based on tensors is not the only possible candidate. There is another related, but different, representational scheme, which also satisfies all the criteria outlined by van Gelder, but which does not suffer from the potential problems of growing representations with increased complexity and constituently structured representations. This is the recursive auto-associative memory.

---

[2] For a description see Rumelhart *et al* [344].

[3] For an alternative representational scheme involving tensors, which do not grow with increased complexity, see Plate [234].

2.2.4.5 *The recursive auto-associative memory*

The recursive auto-associative memory (RAAM) was proposed by Pollack [241, 19]. The RAAM has it in common with tensors that it also involves a technique for associating two vectors (representing the constituents of an expression) via a third vector. Pollack suggested, inspired by the research on n-encoder networks [5], that the hidden layer of a successfully trained auto-associative feed-forward network with two layers of weights can be seen as containing a *reduced description* (see Hinton [143]) of the current input, and that it can be used in further processing.

In contrast to tensors, a RAAM is trained to encode representations for complex expressions by using the backpropagation algorithm [263]. In each processing step the representations for the constituents about to be associated (e.g., *A* and *B*) are presented at the input, and since RAAM is an auto-associative network (figure 2.5) are also used as targets.

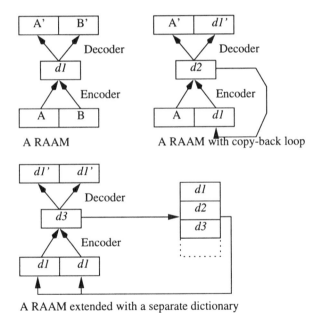

Figure 2.5. Encoding in a RAAM.

The activation on the hidden layer of a RAAM is, as stated earlier, regarded as a reduced description of the pattern combination present at the input. In figure 2.5, *d1* is seen as a reduced description for (*AB*) (at this point a very approximate one, since the training is assumed not to be completed). This reduced description can be combined with other representations, either other reduced descriptions or representations referring to atomic constituents, in further processing (e.g., in generating a reduced description for $(A(AB))$, represented by *d2* in figure 2.5). This is achieved by including a 'copy-back' mechanism from the hidden layer back to an input bank (provided that the slots in question have the same size).

The representations for these encoded tree-structures are often referred to as 'superpositional' representations, since the representations for all the constituents (e.g., sub-branches) of an encoded concept (such as a tree or a sentence) are superposed.

> When two representations are effectively superposed, they become a new item representing both the original contents.... Intuitively, the representations of two distinct items are superposed if they are coextensive—if, in other words, they occupy the same portion of the resources available for representing. ([326], p 43)

If the copy-back mechanism is combined with a separate dictionary then even more complex tree-structures can be stored (figure 2.5). In each processing step the reduced description is saved in the dictionary, which allows the possibility of creating hidden layer representations for expressions like $((AB)(BA))$. This general architecture will of course not only work for binary tree-structures. If three slots are used ternary trees can be stored. This gives us the opportunity to encode representations for complex expressions such as 'Bob knows that John loves Mary', not only by using a dual RAAM, but also by using a ternary RAAM. It is then possible to store the expression as [Bob, knows, [John, loves, Mary]], i.e., first generate the reduced description for 'John loves Mary' and combine that with 'Bob knows...'. One complication is that instead of a compression factor of 2:1 as in the earlier case, we then face a 3:1 compression factor, which means that interference between the stored patterns will increase. A trained RAAM (as was also the case with tensors) seems to possess the ability to represent constituent structure.

Let us now turn to the second criterion to be fulfilled by a functionally compositional scheme; the possibility to decode representations for complex situations (e.g., the tree $(B(AB))$) back into the atomic constituents again. It should be noted that this representational scheme also (like the tensor scheme) typically involves decomposition of superpositional representations into vectors which (only) have a certain resemblance to the original vector, hence the apostrophe for the decoded patterns in figure 2.5.

In a RAAM the decoding of a superpositional representation is achieved by placing the representation about to be decoded at the hidden layer of the RAAM and collecting the representations for the constituents at the output layer. If a decoded representation refers to a complex constituent then it is copied back to the hidden layer for further decomposition. The main problem in this process is to define reliable techniques for classifying decoded representations as referring to atomic or complex constituents. Several techniques can be identified in the literature; comparison between the number of units that are over a certain threshold and the number of active units for the atomic constituents [63], decoding until a special stop-marker is detected [67], decoding a particular number of levels [41], or training the decoder to automatically classify atomic and complex constituents [215]. If an atomic constituent is found, the next problem is to define which token in the represented world the representation refers to. The most popular approach for this classification appears to be to choose the token with the shortest Euclidean distance to the decoded representation. Several simulations have demonstrated that these techniques can be used for stable decomposition of this type of superpositional representation (Niklasson [214], Niklasson and van Gelder [213]).

It appears that RAAMs satisfy van Gelder's definition of a functional compositional scheme, since they possess techniques for both encoding and decoding.[4]   There is, however, an additional component of this issue which van Gelder was neutral towards; *structured* representations.   In a PSS concatenative composition is essential, since it guarantees structured representations (systematicity of representation), which is a pre-requisite for structure sensitive processes (systematicity of inference). Now, so far we have only shown that tensors and RAAMs can be seen as alternatives to classical compositional schemes for representing structure. It remains to be seen whether they generate representations that will allow structure sensitive processes, and if so, what sort of structure these representations have.

2.2.5   Structured representations in a connectionist network

Structured representations are necessary since they are the very foundation on which structure sensitive processes are based; without structured representations no structure sensitive operations are possible.   Let us therefore see how connectionists have shown that their networks can exhibit and explain structure sensitive operations.

The hypothesis of Brousse [51] was that the internal structure of a connectionist network (with two layers of weights) would allow for massive generalization in a combinatorial domain. He showed that a feed-forward auto-associative network (trained on tensor representations) did learn to generalize in the *context independent* domain of Cartesian products. A network trained (for five epochs) on 50 six-letter strings generalized perfectly to about 10 000 novel strings (with another 2.8 million close to being learned).   By using contribution analysis [340], Brousse noted that the hidden layer representations were highly systematic in the sense that similar letter strings generated similar hidden layer representations.   It was also observed that the trained network implemented approximate weight decomposition, resulting in an individual associative network for each letter position. This is interesting since it indicates that the network has detected that the domain is fully context independent, and

---

[4] Please note that Fodor and Pylyshyn's challenge did not demand the ability to encode and decode representations for *infinitely* complex expressions, which is a problem for a RAAM based system.

that influence from neighboring letters could not be used to solve the task.

Sharkey [283] examined the stability of *context dependent* superpositional representations, learned at the hidden layer of a feed-forward network, in the domain of structurally ambiguous sentences. He trained a network to attach the right phrase category to noun-phrases like 'John hit the dog in the market' → (John (hit (dog in market))) and verb-phrases like 'John hit the woman with the stick' → (John (hit woman (with stick))). The input sentences were represented in five slots, by a local representation such as [+john, +hit, +woman, +with, +stick]. In one simulation, a 29–10–1 network was trained to classify 110 sentences into the two classes. After completed training, Sharkey decomposed the hidden layer representations for the sentences 'John hit dog in park' and 'John hit dog on ear' into representations for the individual constituents labeled (ξ). This was achieved by subtracting from the activation at the hidden layer for a complete input sentence the activation at the hidden layer for the same sentence without the input activation of the word about to be decomposed as in (2.1) where f is the sigmoid function applied to the weighted sum of the input constituents

$$\xi_{dog} = f(\text{john.hit.dog.in.park}) - f(\text{john.hit.0.in.park}). \tag{2.1}$$

In a first simulation, the decoded constituent representations were recombined again (by simple addition) and presented at the hidden layer for classification by the hidden-to-output weights. Although the Euclidean distance between the original and the recombined representations were 1.64 (in both cases), the network correctly classified the two recombinations. In a second simulation, the decoded representations for the two sentences were combined with each other, i.e., 'John1 hit1 dog1 on2 ear2' and 'John2 hit2 dog2 in1 park1', but were still correctly classified by the hidden-to-output weights. In a third simulation, a third sentence was introduced into the training set—'John hit dog in ear'. This sentence only differed from the earlier two by one word. When the decoded representations for the constituents were recombined into 'John3 hit3 dog3 in3 park1' and 'John3 hit3 dog3 in3 ear2', the hidden-to-output weights of the network correctly classified both sentences with noun-phrase and verb-phrase attachment respectively. It appears that the representations for the constituents are stable enough to allow syntactic decomposition. Sharkey concluded that distributed representations carry syntactical information, without themselves being syntactically structured. So what is the property that allows these networks to extract and carry syntactical information at the hidden layer?

Pollack [241, 19] showed how his RAAM developed similar hidden representations for members belonging to the same concept type. In one simulation he showed that a RAAM generated similar (in the sense that they seemed to activate a certain hidden unit to the same degree) representations for members of the same phrase-type (such as noun-phrases). In another simulation he showed that words ending with the same letters generated similar hidden layer representations, based on distance in Euclidean space.

2.2.6   Structure sensitive processes in a connectionist network

The models presented in the previous section indicated that the representations formed as a result of learning could have the property needed for systematic processing, namely systematic representations. As stated earlier, a classical account of systematicity of inference relies on structure sensitive operations; in particular a search for certain tokened constituents. It is therefore no problem for a classical symbol system to transform 'John loves Mary' into 'Mary is loved by John', or for that matter 'X loves Y' into 'Y is loved by X'. So, the question is how this can be solved if there are no tokened constituents, or in the words of Fodor and McLaughlin:

> The implication of this difference, from the point of view of the theory of mental processes, is that whereas the Classical constituents of a complex symbol are, ipso facto, available to contribute to the causal consequences of its tokenings—in particular, they are available to provide domains for mental processes—the components of tensor product and superpositional vectors can have no causal status as such. ([105], p 198)

The first simulation investigating the systematicity of inference possible in connectionist networks, using hidden layers as input, was conducted by Pollack [19]. He trained a 48–16–48 feed-forward network to encode/decode representations for relations of the type (LOVED X Y), where the X and Y were chosen from the set JOHN, MARY, PAT, MAN. He then trained a separate network to perform a number of associative inferences (using the hidden layer representations generated by the first network as input and output), such as 'If (LOVED X Y) then (LOVED Y X)'. The latter network was trained on a subset of the possible inferences, but nevertheless processed the remaining correctly. This was the first indication that the representations generated at the hidden layer of a connectionist network had a systematicity of representation that allowed systematic inference.

Chalmers [63] used a similar setup as Pollack, but trained a ternary RAAM to encode/decode 125 sentences of the form 'John loves Michael', and their 125 passive counterparts. After completed training, 75 active/passive sentences were randomly selected and their hidden layer representations were used to train a second feed-forward network to transform an active sentence to a passive sentence.   When tested on the remaining 50 sentences, the network generalized perfectly.   The same result was also obtained when a separate network was trained to transform 75 representations for the passive sentences into the representations for the active.

2.3   A practical demonstration

The empirical studies described in the previous sections all point in the same direction; connectionist networks seem to be structure sensitive, in the sense that they can be trained to exhibit a behavior that appears to be sensitive to the structure of the represented world. Furthermore, RAAMs seem to provide means for generating structured representations, however, an obvious explanation is not available.   Brousse and Smolensky provided insights into the use of tensors for representing a combinatorial domain (with fixed complexity) and structure sensitive transformations. They demonstrated the underlying reasons for the success which relied on unit-localized functions (a kind of structured representation). Pollack and Chalmers demonstrated *performance* that seemed to rely on structure sensitive processes. Based on this one can assume that RAAM is not only capable of representing structure but is also capable of producing structured representations.

However, the actual *performance* of a system can be studied without looking at the internal representations of the system. An *explanation* of the performance must involve the primitives that such systems use to represent the world. We will therefore turn to our own empirical study and see whether it can provide any support to the argument that connectionist representations have structure. We will define a structure sensitive task and investigate the performance of a connectionist network on the task. We will then conduct an investigation of which structural features such representations have.

2.3.1   The environment

The task we set out to study is a classification task that presumably requires *representations that have structure* and the use of a *structure sensitive* classification method.   The task here is to classify representations for tree-structures as being left, right or well balanced, and as having depth 2 or 3 (figure 2.6).

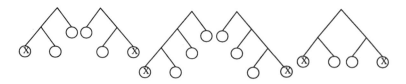

Figure 2.6. Five different tree-structures are used in the study; (i) a left-balanced tree of depth 2, (ii) a right-balanced tree of depth 2, (iii) a left-balanced tree of depth 3, (iv) a right-balanced tree of depth 3 and (v) a well-balanced tree of depth 2. Each tree is, irrespectively of content, terminated with a nil-pattern which is symbolized by 'X'.

The represented world consists of five different tokens, labeled (for convenience) *A, B, C, D* and *E*. The tokens stand in certain relations to each other. These relations are describable as binary trees of either depth 2 or 3. Thus, what is required is that the representing world has means for representing the five tokens combined into binary tree-structures of depth 2 and 3.

A RAAM is used to encode representations for the trees in the domain. The RAAM used here is binary which means that the input and output layers are divided into two slots. Also, it is used to encode tree-structures with different balance, which means that it must be possible to copy the hidden pattern to any of the two input and output slots.

The leaves of the trees are set to patterns identifying tokens. Five different tokens are used in this study, each is represented by a unique pattern which has two active elements (both set to 1) and is orthogonal to all others (table 2.1). Subsequently, ten units are used to represent one token.

Table 2.1. Patterns representing the tokens.

A	1	0	0	0	0	1	0	0	0	0
B	0	1	0	0	0	0	1	0	0	0
C	0	0	1	0	0	0	0	1	0	0
D	0	0	0	1	0	0	0	0	1	0
E	0	0	0	0	1	0	0	0	0	1
nil	0	0	0	0	0	0	0	0	0	0

2.3.2 Performance on RAAM-generated representations

Five RAAMs, initialized with different weight-settings (generated randomly from different seeds), were trained to encode and decode all possible trees of the five structures, with the five tokens appearing in all positions. With a nil-pattern marking the end of the tree, this adds up to 325 different trees.

Each RAAM was trained using backpropagation for 10 000 cycles, where each cycle involved iterating through all of the 325 trees. 10 000 cycles were sufficient to reach a stable summed squared error for all networks. The learning rate was 0.01 and the momentum term was 0.5.

The hidden layer representations, for the 325 tree-structures, were then subjected to three types of process: recursive decoding, and holistic depth and balance classification. The latter two tasks demanded that two separate feed-forward networks were trained to perform the classifications on the generated hidden layer representations. Earlier demonstrations [43] had shown that RAAMs encoded structural features which allowed holistic classification to some degree.

Table 2.2. Errors in the RAAM decoder. The first column contains the number of errors for decoding the token in the last leaf (i.e., the leaf presented in the last encoding step in the RAAM), the second column contains the errors for the second-to-last leaf. The third column contains the errors for decoding the terminator (i.e., the nil-pattern marking the end of the tree) and the fourth column contains either the decoding error for the second nil-pattern (if its a well-balanced tree) or the error for decoding the initial token used. Note that there is an overlap between this number and the errors for the second last token in the trees of depth 2. The last column sums up the errors made. The maximum number of errors that can be made is 1250.

Run No	Last	2nd last	Terminator	2nd term or initial	Total
1	0	48	51	204	288
2	1	5	27	176	209
3	0	13	69	176	254
4	0	3	52	179	231
5	0	0	40	175	215

2.3.2.1 Decoding representations

The decoding of the generated superpositional representations relied on prior knowledge about the stored tree-structure, in order to decode the correct balance and number of levels.[5] For each decoded representation referring to an atomic token, the token with the shortest Euclidean distance to the decoded representation was chosen. The RAAMs did not correctly decode all generated patterns into their atomic tokens. The errors (table 2.2) were made almost exclusively among the trees with depth 3, at depth 3.

Decreasing the number of units used for representing trees deteriorated decoding performance. Increasing the number to 20 units resulted in perfect decoding performance. No errors were made in any of the five runs. However, we are here not primarily concerned with under which conditions the RAAM decodes correctly. Instead we intend to investigate whether or not RAAM-generated patterns have structural features. Therefore we used the ten-hidden-unit setup for our further analyses.

2.3.2.2 Classifying structural features

The hypothesis that a separate network could extract potential structural features encoded in the hidden layer of a RAAM was then tested. Two separate networks were trained to classify the patterns produced for the trees into different classes depending on two structural features, i.e., depth and balance. The training set

---

[5] However, the results presented in the next section show that this need not be the case. Instead, the network could be *trained* to handle this automatically.

for the classification tasks was composed of 180 of the 325 trees, with 145 kept aside for testing generalization performance. Rather arbitrarily, the training set was selected such that no trees contained more than one instance of each token.

Of the 325 trees generated by the RAAMs, 75 were of depth 2 and 250 were of depth 3. According to the above criteria for being part of the training set, 15 of the depth 2 trees and 130 of the depth 3 trees were not used during training of a depth classifier. A separate connectionist network without hidden units and two output units was trained by backpropagation to classify the depth for 180 of the RAAM-generated patterns. Each class within the task was allocated one separate output unit. The learning rate was 0.01 and the momentum was 0.5. The run was repeated five times, once for every run of the RAAM. After training, the whole 325-tree domain was presented in order to evaluate the performance of the classifier (table 2.3).

150 of the 325 trees generated by the RAAMs were left balanced, subsequently 150 were right balanced and 25 were well balanced. Of these, five of the well balanced, 70 of the left balanced and, subsequently, 70 of the right balanced were set aside and used for testing and evaluating the generalization performance of a balance classifier. A connectionist network without hidden units and with three output units was trained by backpropagation to classify the balance for 180 of the RAAM-generated patterns. Each class within the task was allocated one separate output unit. The learning rate was 0.01 and the momentum was 0.5. As for the depth classification, the learning was performed five times; the results when the whole domain was presented to the trained classifier are shown in table 2.3.

Table 2.3. The errors in the five runs of the classification networks for depth and balance. The total number of trees tested was 325.

Run No	Depth	Balance
1	0	16
2	7	0
3	14	0
4	0	0
5	6	0

The classification networks completely learned the training set and generalized to the 145 patterns not included in the training. The results indicate that the RAAM generates representations that contain structural features and that a separate connectionist network can be sensitive to such structural features. It seems that the two competing systems (NSS and PSS) can be, at least in principle (see run 4), equivalent at a computational level in a structural domain.

Table 2.4. The 70 trees that are used in the demonstrations.

1	((nA)A)	2	(A(An))	3	((nA)(An))	4	((nA)B)
5	(B(An))	6	((nB)(An))	7	((nA)C)	8	(C(An))
9	((nC)(An))	10	((nA)D)	11	(D(An))	12	((nD)(An))
13	((nA)E)	14	(E(An))	15	((nE)(An))	16	((nB)A)
17	(A(Bn))	18	((nA)(Bn))	19	((nC)A)	20	(A(Cn))
21	((nA)(Cn))	22	((nD)A)	23	(A(Dn))	24	((nA)(Dn))
25	((nE)A)	26	(A(En))	27	((nA)(En))	28	(((nA)A)A)
29	(A(A(An)))	30	(((nA)A)B)	31	(B(A(An)))	32	(((nA)A)C)
33	(C(A(An)))	34	(((nA)A)D)	35	(D(A(An)))	36	(((nA)A)E)
37	(E(A(An)))	38	(((nA)B)A)	39	(A(B(An)))	40	(((nA)B)B)
41	(B(B(An)))	42	(((nA)B)C)	43	(C(B(An)))	44	(((nA)B)D)
45	(D(B(An)))	46	(((nA)B)E)	47	(E(B(An)))	48	(((nA)C)A)
49	(A(C(An)))	50	(((nA)C)B)	51	(B(C(An)))	52	(((nA)C)C)
53	(C(C(An)))	54	(((nA)C)D)	55	(D(C(An)))	56	(((nA)C)E)
57	(E(C(An)))	58	(((nA)D)A)	59	(A(D(An)))	60	(((nA)D)B)
61	(B(D(An)))	62	(((nA)D)C)	63	(C(D(An)))	64	(((nA)D)D)
65	(D(D(An)))	66	(((nA)D)E)	67	(E(D(An)))	68	(((nA)E)A)
69	(A(E(An)))	70	(((nA)E)B)				

However, there is nothing in the above results that provides substantial evidence for the assumption that the hidden layer representations used *have* structure.

2.3.3 Detecting structured representations in RAAM

In order to highlight the features that encode the structural information, a number of analyses were conducted on individual hidden units, the complete set of hidden units and the weight sets.

2.3.3.1 Bands of activation

Inspections reveal a number of interesting features. By plotting all patterns unit-wise several *bands* of activity can be found,[6] see figure 2.7 and table 2.4 where 70 of the 325 representations for tree-structures have been studied.[7] Some units only produce outputs of a limited variety, e.g., hidden unit 2 of the RAAM in run 1 only produces an output of approximately 0.1, 0.2 and 0.9. It seems that some information is not completely distributed since it is possible to detect some coherence between the bands and the periodicity of certain features of the trees. For example, the same unit (unit 2) is one source for determining

---

[6] This type of analysis is similar to that performed by Berkeley *et al* [39].
[7] The trees were selected according to the 70 first where the token *A* appeared at least once. This resulted in a fairly representative subset of the representations.

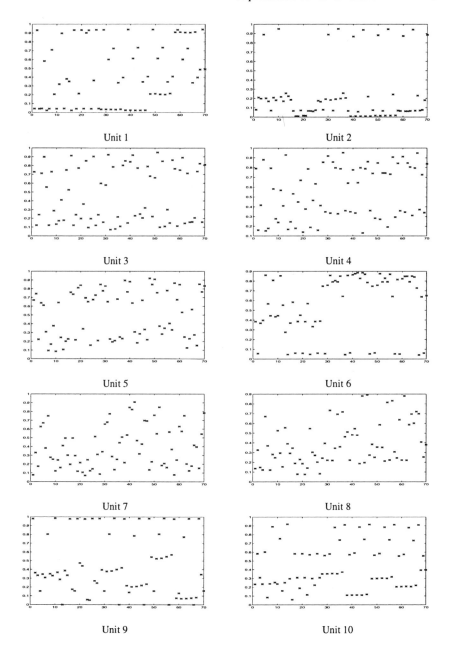

Figure 2.7. The activity for units 1 to 10 in the representations for 70 of the 325 tree-structures. The activity is plotted on the *y*-axis pattern-wise (*x*-axis). The plots are derived from the first of the five runs.

whether token *A* appears as the last leaf in a left-balanced tree (the value is always approximately 0.1). For unit 6 it is quite obvious where the border between trees with depth 2 and 3 lies. At pattern 28 (see the *x*-axis) patterns for representing trees with depth 3 start (the output is almost limited to the band at 0.8). Before that only trees with depth 2 occur.

There are some interesting bands for unit 1. At a closer look they correspond to all left-balanced trees in the domain. Whenever the tree has *A* or *B* as second-to-last leaf in a left-balanced tree unit 1 has value 0.04. Whenever token *C* occurs in the same position unit 1 has value 0.2. When token *D* is present at the same position unit 1 has activation 0.9. *E* is more vague but the corresponding value is around 0.4.

Many other observations can be made in this way, but we restrict ourselves in this context to note some cases and argue that the information contained in the representations for the tree-structures is not necessarily distributed. It is not, however, localized in the same way as for units in a localist network. The interpretation for a unit is not restricted to a particular concatenative feature in the same sense as with symbolic expressions (and localist representations). Therefore we are not sensitive to the critique of Fodor and Pylyshyn that any connectionist solution can be viewed as an implementation of its symbolic counterpart. It is believed that the information extractable from this type of analysis may complement the information we can extract using other methods.

In a similar way we argue that not all information is localized. There are cases when bands did not appear. In five of the ten units bands were found. In the other units no bands were detected. Also, a single unit value does not identify a feature by itself. Furthermore, the single value may be involved in the identification of different features depending on the band it belongs to. To fulfill the classification tasks successfully a holistic process, which operates on the whole set of values, is required.

2.3.3.2 *Applying hierarchical cluster analysis*

A different method, called *hierarchical cluster analysis* (HCA), is based on closeness measures between vectors, e.g., Euclidean distance (for an introduction to HCA, see Chapter 1 of [1]).

The question now is if we can extract knowledge of which structural features are available in the patterns generated by RAAM. We are not explicitly interested in the output classes of the RAAM, merely the structure imposed during learning and its implications for the hidden unit activity.

The Euclidean distance (ED) approach pays no special attention to intra-dimensional information (e.g., bands). The observations made in the previous section suggest that such measures might not identify features as distinctively as intra-dimensional measures. One measure that could be put forward as an alternative to ED is the city-block distance. This measure favors vectors which share values in specific dimensions. However, both these measures generated very similar dendrograms when applied to the representations formed in our RAAMs.

A hierarchical cluster analysis was performed on 70 of the 325 generated patterns (see table 2.4) to demonstrate the spatial relatedness between patterns in representational space. The patterns generated in the first run were used. A Euclidean distance metric was used and average linkage, i.e., the average distance for all the vectors in the cluster, was used for merging clusters. The dendrogram is shown in figure 2.8.

Ten clusters stand out in the diagram. These are clearly based on having a particular token as the last leaf of the tree. Within these clusters there are tendencies such that whenever a particular token appears twice in the second-to-last leaf these two are close. Well balanced trees also stand out.

Structural features are harder to detect. Consider the ten clusters having a particular token in the last leaf in a left- or right-balanced tree and the cluster of well-balanced trees. They can be thought of as sub-classes of the balance classes. In this sense, it is not hard to understand that the balance feature is available to the classification system.

The depth feature is not obvious in the cluster diagrams. The main feature in the cluster diagrams is the last leaf in the tree, independently of depth. Within these clusters representations for trees of the same depth tend to cluster. However, this is not nearly as evident as indicated by the results from the classification networks.

2.3.4 Explaining the presence of structure with decision hyperplanes

Each receptive unit in a network implements a decision hyperplane in terms of the weights that connect to it. For detailed descriptions see [342, 244, 341]. Ideally, the output function is a Heaviside function which discretely discriminates between 'on' and 'off'. This condition can be relaxed. Many output functions, such as the logistic function used in this study, are monotonic but continuous. This need not be a problem in itself since there is still a value that can be used to determine where the border between 'on' and 'off' lies. To ensure that only one 'switch' between 'on' and 'off' occurs, the output function needs to be monotonically increasing (or decreasing). Hence, the selected output function determines the level of summed activation that is needed in order for the unit to be 'on' or 'off'. For the logistic function,

$$f(net) = \frac{1}{1 + \exp(-net)}$$

the possible output ranges between 0.0 and 1.0. The obvious midpoint is 0.5. To produce 0.5 a *net* input value to a unit of 0.0 is needed. Thus, whenever the summed input exceeds 0.0 the unit is turned 'on' and whenever it is below 0.0 the unit goes 'off'. Hence, the hyperplane is described by the summation function when it equals 0.0 ($w_1 i_1 + w_2 i_2 + \cdots + w_n i_n + w_\theta = 0.0$).

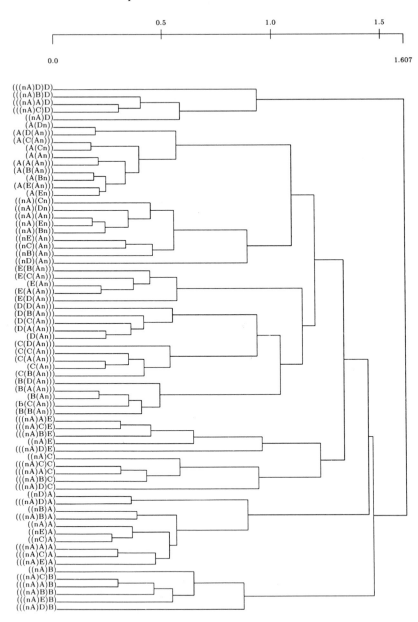

Figure 2.8. The hierarchical cluster analysis dendrogram using Euclidean distance with average linkage for the first run. Note that not all patterns are used. The patterns were selected objectively.

The natural extension here is to describe this as a formal rule of the form, if $=(w_1 i_1 + w_2 i_2 + \cdots + w_n i_n + w_\theta > 0.0)$ then what the unit is representing is true, else what the unit is representing is false.

We have shown that some structural features are evident either by observing activity at the unit level or by performing cluster analyses. If these features appear in general (as indicated by the experiment) how can their occurrence be explained? There is an approach that makes use of hyperplane analysis that sheds some light on this. This approach is used by Bodén [42] to explain how RAAM can be used to generate type/token representations.

This approach relies on the fact that RAAM is auto-associative and the assumption that the learning error for the RAAM is close to zero. For any input in the training set, where some parts are known in advance, there exists a constrained region in which the hidden activity always will end up. This fact relates to the learning of hyperplanes (i.e., weight configurations) in a second network (e.g., a classification network that classifies a pattern representing a tree into a set of classes), which can detect and make use of these regions. If the second network does this, it will generalize to any input filtered through the RAAM.

The underlying principle behind the identification of regions in space relies on the availability of partial input/output activity. For a binary tree $((AB)C)$ a binary RAAM is required. To encode a representation for the tree, the pattern for A is first combined with the pattern for B at the input layer. Assuming that the RAAM has a learning error of zero, the two patterns are replicated at the corresponding output slots. Knowing only one of the tokens, such as B in (xB), the pattern for B can be used to determine on which sides of the hyperplanes in the second half of the hidden-to-output layer the hidden activity must fall. Similarly, the constrained hidden region for trees matching (Ax) is identified by going through all values in the A pattern to determine if the hyperplane of each unit is turning the unit on or off. This information is then used to decide on which side the hidden activity must fall if this output is to be produced.

In figure 2.9 the RAAM is binary and has two units in each slot. The first output slot determines two hyperplanes cutting the hidden unit space; H1 and H2. Consequently, the second output slot determines H3 and H4. If the activation in the first slot is known, as for (Ax), the pattern representing A can be used to determine the region for all complex representations matching that specification. Imagine that the representation for A is [0.0, 1.0]. This means that in order for the first output slot to be [0.0, 1.0] (and we know that it should since successful learning is assumed) the hidden activity must be on the 'off' side of H1 and on the 'on' side of H2.

Combining (AB) with C works in the same way. The presence of such regions explains why patterns for representing trees cluster with respect to the token in the last leaf. They all share the values in one of the output slots. This means they will be contained in a region in the ten-dimensional hidden space in terms of ten hyperplanes. Furthermore, an incomplete tree, such as (xB), can

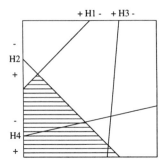

Figure 2.9. Regions in hidden space are determined by expected output activations. The 'on' and 'off' regions are marked + and −, respectively.

be combined with C by projecting the hidden region for (xB) on to the first slot of both the output and input. Without going into detail, this means further information can be derived on how regions in hidden space are used to convey information about trees of poorly specified constituents.

2.3.5   Explaining the performance of the classifiers

Since we have identified a few bands among the patterns for representing tree-structures the weights of the classifiers also provide more knowledge on how these bands are used. A Hinton diagram[8] reveals the importance of a number of input features for a particular task. This type of analysis is particularly useful in networks with only one layer of weights. The main idea is to plot boxes with sizes that correspond to the magnitudes of the weights. Positive weights are distinguished from negative weights by the color of the box. The size of the box indicates the importance of the input unit for the output unit. Hinton diagrams were generated for both the depth and balance classification task. Weights from the first run were used.

In the Hinton diagram for the depth-classifier (figure 2.10) none of the inputs stand out. A few things deserve to be noted. First, the weights are the same for both output units but of opposite polarity. Second, the weights are fairly small and none of them indicate that any of the input units have any special significance. The representational features that identify the two classes seem distributed with some importance focused on elements 4, 8, 9 and 10. Since they complement each other it is hard to detect this by visual inspection of the unit band analysis.

In the Hinton diagram for the balance classifier in figure 2.11, a few units stand out. Although the classification network makes the decisions on the basis of all the units it provides some guidance on where to find the most striking representational features for the generated trees. Considering the classification

---

[8] For substantial examples and further explanations see Hinton [143].

Figure 2.10. Hinton diagram that indicates which input features are relevant for the depth classification task. Output 1 corresponds to depth 2, and output 2 corresponds to depth 3. White boxes indicate negative weights between the input and output units; black boxes indicate positive weights. The size of the box corresponds to the magnitude of the weight.

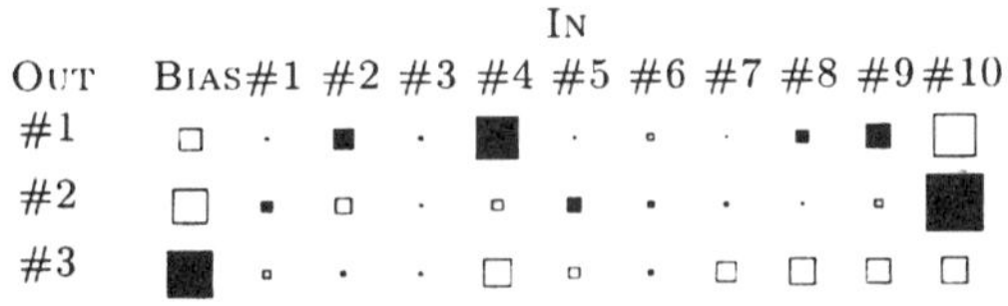

Figure 2.11. Hinton diagram that indicates which input features are relevant for the balance classification task. Output 1 corresponds to left-balanced, output 2 corresponds to right-balanced and output 3 corresponds to well-balanced trees.

network units, 4, 9 and 10 seem fairly important for judging whether a particular representation is for a left-, right- or well-balanced tree. Considering only the representations (not the classification network itself) two units are striking. A unit band analysis was performed on unit 4 and 10 where they were plotted separately depending on their class (figure 2.12). Note that all patterns were plotted. Referring to the unit 10 plots, it is possible to distinguish between all right-balanced trees and all others by just looking at the activity of this unit. Whenever the activity of this unit is above 0.5 it is a right-balanced tree that is being represented. However, there is an overlap between activities for left- and well-balanced trees. Again by looking at the Hinton diagram in figure 2.11 it is noted that unit 4 may distinguish between left- and well-balanced trees. The unit activity was plotted with respect to the same classes, see figure 2.12. It is not as obvious as for unit 10 but the plots indicate that the unit reveals information that can be used to discriminate between the two classes of patterns.

The obvious comment here is that surely the representations are structured; they have a *spatial structure*. Many structural features can be detected, although sometimes with help from a different network. Most of the time, however, they are distributed over many units. This should not be interpreted as a weakness but as an advantage to accommodate much more information than simple syntactical features.

It is quite easy to show that the classifier networks do not memorize the training set, instead they generalize to accommodate the samples outside the

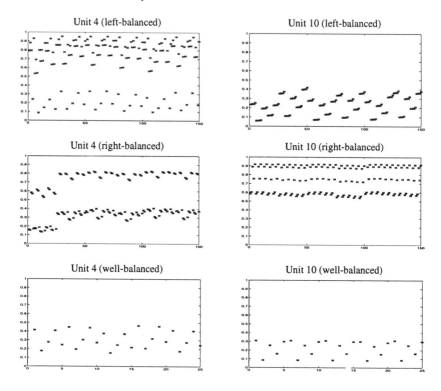

Figure 2.12. The activity of units 4 and 10 plotted with respect to their balance classification.

training set. One way of demonstrating this would be to plot the decision hyperplanes of the classifier networks to show how simple the mappings really are. Obviously this is not possible for a ten-dimensional input. The hyperplanes can, however, be described as linear functions. Merely the fact that they are linear and that there are only two or three of them indicates how simple the mappings really are. Following the ideas expressed earlier we outline the rules implemented by the classifiers by equations depth and balance, below.

function depth $(x_1, x_2, \ldots, x_{10})$ returns $\{2, 3\}$
if $(2.0x_1 + 1.9x_2 - 2.2x_3 + 10.3x_4 + 5.9x_5 - 2.2x_6$
$\quad + 6.1x_7 + 11.6x_8 + 7.7x_9 + 9.8x_{10} - 22.2 > 0.0)$
$\quad$ then return 2
else return 3

Note that the same hyperplane is used to distinguish between both classes. That is, they cut the space at the same position. Obviously the two hyperplanes have opposite polarity.

function balance $(x_1, x_2, \ldots, x_{10})$ returns {left, right, well}
if $(-0.9x_1 + 4.7x_2 - 1.0x_3 + 9.9x_4 - 0.7x_5 - 1.7x_6$
$\quad + 0.6x_7 + 3.2x_8 + 5.7x_9 - 9.8x_{10} - 5.2 > 0.0)$
then return left
if $(2.7x_1 - 3.8x_2 - 0.8x_3 - 2.7x_4 + 3.4x_5 + 1.6x_6$
$\quad + 1.0x_7 + 0.7x_8 - 2.0x_9 + 13.0x_{10} - 8.0 > 0.0)$
then return right
if $(-2.0x_1 - 1.2x_2 + 0.9x_3 - 6.4x_4 - 2.7x_5 + 1.3x_6$
$\quad + 4.8x_7 - 6.1x_8 - 5.7x_9 - 6.1x_{10} + 10.6 > 0.0)$
then return well

The simplicity with which the classification networks handled the tasks only goes to show that the evaluated structural features must be present in the inputs and not generated or developed in the classification networks. Furthermore, it should be noted that the RAAM generated these structural features without any external influence.

2.4 Discussion and conclusions

We can now return to the issues raised in the beginning of this chapter. We set out to study two computationally equivalent systems, i.e., two systems that solved the same task, but in totally different manners. There are basically three sides to this issue; performance, and the aspects of the representing world that are doing the modeling, i.e., the type of representation and process used by the system.

(i)   We have shown that an NSS can successfully perform classification of the representing world, based on structural features of the represented world.

(ii) Our claim that the NSS can not only represent structure, but also have structure (i.e., that the representing world has properties which can be explicitly formalized) was supported by the results from a number of analyses. We showed how a RAAM generates hidden layer representations, in the form of locations in a multi-dimensional space, such that representations for related expressions were grouped into the form of regions (a spatial structure), constrained by the hidden-to-output hyperplanes formed as a result of learning.

(iii) The type of process possible in a system is constrained by the choice of representing world. The basis for systematic processing in a PSS is that related expressions have representations with similar constituent, or syntactic, structure. If that is the case, the processes can operate on the form of the representations. In an NSS the representing world has a different form. Related expressions are located to related areas in a multi-dimensional representational space. In the same way that a symbolic representing world can be multi-purpose, so also can a non-symbolic representing world. Separate processes, decoding or different types of classification, can constrain the actual locations for the expressions into different regions, depending on the task of the process. This means that *what the NSS is doing is to manipulate spatial form*, compared to the *manipulation* of *syntactic form* done by the PSS.

The intention of this chapter has been to present two systems which perform the same task, but use different techniques to do it. The reasons why we feel that it is important to show that a connectionist system can indeed represent structure as well as have structure, is that we feel that the view of connectionist systems as mere implementations of symbol systems is wrong. The demonstrations that we have given here have shown that connectionists can supply an alternative to the structure sensitivity used in symbol systems. This alternative is based on processes sensitive to different *regions of representational space*, constrained by weights. If this approach is accepted then there appear to exist certain systems which can exhibit and explain a kind of systematicity of thought, but which do not naturally line up with the three-level classical cognitive architecture. There are, for instance, no 'symbols' in superpositional representations available for structure sensitive processes.

The system we have presented might be accused of being a hybrid, since we use a separate dictionary to store the superpositional representations. We feel that such criticism, although perhaps justified, in no way compromises the claim that the system is not an implementation of a PSS.

It appears that the number of representational level explanations (i.e., what the system is actually doing) of cognitive phenomena is not limited to those possible in a PSS, but could also include a number of explanations available in an NSS. Which of the explanations best describes individual cognitive phenomena remains an open question.

Chapter 3

Chaos, Dynamics and Computational Power in Biologically Plausible Neural Networks

Robert W Kentridge
University of Durham, UK
robert.kentridge@dur.ac.uk

3.1 Introduction

The functioning of mind and brain is sometimes characterized in terms of symbol manipulation and sometimes by some sort of neural network. Nevertheless, no-one would propose that the brain consists of separate symbolic and neural processors. The brain is clearly fundamentally neural yet its style of computation often appears to be symbolic. How can we produce symbolic interpretations of neural networks' behaviour and when might such a level of description become appropriate? In this chapter I use an algorithm developed by Crutchfield and Young to produce symbolic descriptions of the behaviour of a simple neural-network model of the cerebral cortex. I explore some of the conditions under which the computational power of the symbolic interpretation of the network's behaviour is maximized, relating this computational power to the computational demands of cognitive processes. Finally, I consider the implications that the study of this toy problem has for symbolic approaches to artificial intelligence.

3.2 Symbolic systems

Symbolic models of cognition are based on collections of rules which determine how a model responds as it encounters stimuli. The major assumption made is that stimuli and the internal states of the model can be represented by tokens which are arbitrary placeholders—they do not reflect any of the qualities of the

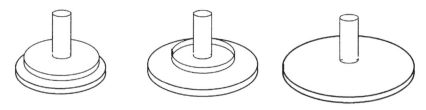

Figure 3.1. The Tower of Hanoi Puzzle. The aim is to move the pyramid of discs from one peg to another, using the third peg as needed, one disc at a time. The caveat in the rules is that at no time may a larger disc be placed on top of a smaller disc.

stimuli or states themselves. The great power of these models is that the rules which operate on these symbols can be applied generally, as we see when solving problems this way. To solve the Tower of Hanoi problem (see figure 3.1) we can use the symbolic algorithm shown below.

In general to move a tower of height n from a source location src to a destination location dst using the third location tmp on the way:

If tower has height 1 then move disc from src to dst, else:

- **1 move tower height $n - 1$ from src to tmp using dst**
- **2 move disc from src to dst**
- **3 move tower height $n - 1$ from tmp to dst using src**

If it also defines what happens to the height of towers when we move a disc then the symbolic description above describes how to solve the Tower of Hanoi problem for *any* size of tower. Two key features lead to this power—first, the way in which symbols act as arbitrary placeholders which can stand for any specific instance of towers and second, the extensibility of the algorithm itself to towers of all sizes.

The aim of symbolic computational approaches to the study of cognition has been to apply similar techniques to the description of psychological and everyday tasks (e.g., object recognition or buying a hamburger in Macdonald's). This endeavour has, however, been fraught with problems.

3.3   The problem of knowledge explosion in symbol systems

The statement 'if x can fly **and** x has wings **then** x is a bird' seems a reasonable first attempt at a symbolic rule for bird identification, however, it identifies Boeing 707s as birds and fails to identify penguins, chickens, stuffed eagles, newly hatched starlings and so on as birds. Extra rules could be added to deal with all these eventualities, however, as the categorizations demanded from the system become more complex then the number of rules required becomes quite unmanageable. Attempts to deal with this combinatorial explosion of rules by encoding default background knowledge using further symbolic rules such as *scripts* [269] and *frames* [196] are only partially successful—they tend to be inflexible in their reliance on prior knowledge supplied by the system's builder.

3.4 Neural networks

Neural networks appear to offer an alternative approach which might avoid combinatorial explosion when dealing with complex categorizations. Neural networks consist of a large number of neuron-like elements connected together by links through which the activity of one unit influences the activities of those it is connected to. Entities are represented in neural networks as patterns of activation in these units. The strength of the connections between units determines how the pattern of activity in the network changes in response to an input stimulus. Depending on the arrangement of connections the effect of an external input can either propagate through the network in a single pass, or, if some of the connections feed back into the net, its effects can develop over time (perhaps forever). In this way the pattern of activity produced in some of the units by an externally applied stimulus, together, perhaps, with the initial state of the network, evolve to produce a new pattern of activity which represents the network's interpretation of the stimulus. Given an appropriate method of adjusting the strengths of the connections between units a neural net can be made to classify stimuli and produce useful outputs even if the stimulus is incomplete or quite novel.

Many methods of adjusting connection strengths in neural networks have been devised and have been used to solve real problems. This appears to be a very promising situation; however, more detailed consideration of the way in which knowledge is represented in neural networks reveals potentially serious problems.

3.5 Representing rules and relations in neural networks

There are two broad classes of neural networks, those in which each of the units in the net stands for some quality (e.g., redness, having feathers) and those in which some input units stand for qualities of the stimulus and some output units stand for qualities in the system's encoding of the stimulus (e.g., can fly, is a connected figure etc), but most units are hidden and, although the relationship between input and output depends on them, they have no explicit individual meaning—their response to particular stimuli is determined by the algorithm used to construct appropriate connections in the net. In both cases, however, the representation of a stimulus in the network, unlike its representation in a symbolic system, is not arbitrary—it is dependent, albeit indirectly, on some qualities of the stimulus itself reflected in the coding of the stimulus in the input to the network. As representations are not arbitrary and are distributed over the whole network it is not possible to distinguish between a state of the network which represents the presence of two separate objects and one which represents a single complex object. The consequences of this are very serious— there is not only a problem representing multiple objects, it also follows that it is equally hard to represent rules. First, we cannot represent relations between two specific objects so we cannot represent rules pertaining to those objects. Second, we cannot represent arbitrary objects with particular properties to which specific instances of objects can temporarily be associated so we cannot make use of the power of arbitrary placeholders that is used in symbolic algorithms. As Fodor and Pylyshyn [106] argue, models of thinking require systematic representation of knowledge at a very minimum—that is, knowledge represented in a manner which allows generalizable relations to be represented. This does not appear to be the case for most neural networks.

3.6 Time and binding

One obvious method of allowing multiple representation to exist in a net, which is the first step towards encoding relations and rules, is to separate representations in time. This principle can be applied either sequentially, so that representations of items and relations follow one another, or in parallel so that representations of multiple items and relations exist together in the network but are distinguished on the basis of the temporal characteristics of their encoding (for example in terms of the relative phase or oscillation frequency of units involved in particular representations). In general, rule representation has been the explicit aim of networks using parallel temporal coding schemes (e.g., [284] [285] [329]), while it emerges and may not even be recognized in networks which identify structures over time. These networks are, however, capable, in principle, of encoding relations between items as well.

3.7 Computational power and network dynamics

Temporal separation may allow multiple objects to be represented in neural networks, but what of the other property of symbolic algorithms which gives them such power—their extensibility over many different instances of problems? One way of characterizing the range of problems which algorithms can deal with is in terms of the complexity of the regularities that occur in them—their formal computational structure. Chomsky [66] defined a hierarchical classification of computational power in terms of the types of grammar (or their equivalent automata) necessary to generate (or parse) a given behaviour. The dependences between the states produced by a process can be characterized either in terms of a class of grammar or an equivalent formal machine:

- **Type 0: Unrestricted Turing Machine FSA with infinite tape**
- **Type 1: Context-Sensitive Linear-Bounded Automaton FSA with finite tape**
- **Type 2: Context-Free Pushdown-Stack Automaton FSA with stack**
- **Type 3: Regular Finite-State Automaton**

This is a strict hierarchy; no grammar in this hierarchy can accept (or produce) behaviour only describable by a higher-level grammar. The qualitative nature of this classification allows us to characterize the computational properties of neural networks obeying different principles of operation regardless of the size or precise implementation of those networks.

It has been shown that purely feedforward networks with at least one hidden layer of units are, in principle, capable of approximating any function in their mapping of input onto output states [147]. This corresponds to the highest level of the Chomsky hierarchy—unrestricted grammar. I would like, however, to draw a distinction between the computation performed by a neural network as it responds to stimuli and the computation performed by the combined processes of training the network and pre-processing the stimuli combined with the network's response to those stimuli. Although a feedforward network may behave as a universal function approximator, the network's response to stimuli is essentially just a matter of table look-up (which can be conceived of as a very simple regular grammar in which a single transition is made from each starting state to a corresponding terminal state). There may, however, be regularities in the task which are structured in time but which vary too quickly for such approximations to succeed. If the time scale of regularities in the task to be learned and of the network's behaviour are similar but the time scale of adaptation is slow then a table look-up approximation will inevitably fail whereas equally slow adaptation of behaviour which reflects the real computational structure of the task can still succeed.

Having put computation in networks with no feedback, and hence no dynamics, to one side, let us now consider the behaviour of dynamic networks in a computational framework. The behaviour of networks with dynamics is not just dependent on the stimuli to which they are exposed at any instant, but also on their prior state. If we wish to make inferences about the computational power of networks it is important to establish how far this dependence of responses on the processing of prior stimuli extends back in time. There is a clear computational distinction between systems in which the transitions between states depends on current state and input—finite-state automata—and those in which state transitions depend on current input and an arbitrarily long history of prior states—pushdown-stack automata, linear-bounded automata, Turing machines etc (the distinctions between these latter types derive from constraints on the construction of the state histories they can use). What implication does this argument have for common approaches to information processing with dynamic neural networks?

The most common method of processing information with dynamic neural networks is to encode target states as point attractors in the network's dynamics. From any given starting state, the state of the network will therefore evolve towards the attractor state whose basin of attraction the starting state happened to be in. The rules used to determine where attractors are formed and the addition of noise and other constraints allow this general scheme to be used in a wide variety of problems. The normal way of using these networks is of minimal computational power. The transition from an arbitrary starting state to an encoded attractor state is, once again, simply a mapping. If, however, we ask the question of how a subsequent stimulus might affect the network the situation becomes a little more interesting. If this new stimulus perturbs the states of some, but not all, of the neurons in the network then the basin of attraction into which the network consequentially falls depends both on the new stimulus and the network's prior state. The transitions between basins of attraction (the informationally significant subdivisions of the network's states) depend on the history of the network. This dependence is, however, only the single-step dependence of a finite-state machine, not the arbitrarily long time dependence of more powerful stack- or tape-machines. This becomes clear when we consider the nature of attractors. The reason that attractors are useful information processing tools is that their use allows a system to selectively throw away irrelevant information. All of the different states of a system within one basin of attraction are reduced to a single attracting state. Once the network has reached one of these attracting states no information remains which can distinguish where in that state's basin of attraction the network had previously been. When a system makes a transition to a new attractor all information of its prior states is therefore lost. The direct dependence of the response of an attractor network to a stimulus cannot therefore extend beyond it current state. The same argument applies to networks which support multiple limit-cycle or chaotic attractors—the point is that once the system has converged onto one of a set of attractors it cannot directly recover information about which other members of the set it has visited or when it did so.

Any system using attractors to represent the informationally significant subdivisions of its state space does not appear capable of supporting computation above the level of finite-state automata. Nevertheless, physical dynamic systems which have greater computational power clearly exist and some natural examples appear to have neural-network processors! How can their dynamics be being harnessed for computation if not in terms of a sequence of attractor transitions?

Much has been made in recent work on information processing in complex dynamical systems of the importance of the edge of chaos [224, 78]. Phenomenologically, the interesting thing about physical systems undergoing slow, second-order or critical phase transitions (which are at the edge of chaos) is that their behaviour depends on interactions between their components extending over arbitrary distances of space and time [84, 321]. This is very interesting in terms of the computational consequences of dynamics since it implies that under these conditions computation of greater power than finite-state automata may occur.

3.8 Experiments on network dynamics and computational power

It would clearly be interesting to examine the computational power of networks with critical dynamics. Most artificial neural networks are not constructed with these dynamics in mind. On the other hand, both the preceding argument and some evidence from electrophysiological recordings [112, 336, 160], suggest that real brains may well support such dynamics. I have therefore studied these phenomena in network models which are closely based on the anatomy and physiology of the cerebral cortex.

3.9 A simple, biologically plausible, neural-network model

Ignoring the layered third dimension of cortical structure we can consider the cortex as a two-dimensional sheet of cells in which there is a spatially dependent organization of connectivity between neurons [46, 181]. In the model the probability of a connection existing between neurons is dependent on a Gaussian of the distance between them. Figure 3.2 shows the input connections to a small percentage of neurons in a typical model network.

Although some very sophisticated models of single-neuron physiology have been produced [183] our aim here is to reproduce some of the representative non-linearities of neurons as simply as possible, since we need to simulate networks comprised of many thousands of cells. The model neuron used in all the simulations described here was therefore a spiking leaky integrator which also exhibited absolute refractoriness after firing, described formally as follows:

$$\text{if} \quad v_j(t) > \text{th}_j \qquad x_j(t) = \text{ap}_j \quad \text{and} \quad v_j(t+1+r) = \Sigma w_{ij} x_i(t+r)$$

$$\text{if} \quad v_j(t) \leq \text{th}_j \qquad x_j(t) = 0 \quad \text{and} \quad v_j(t+1) = \Sigma w_{ij} x_i(t) + d_j v_j(t)$$

where $v_j(t)$ represents the membrane potential of neuron j at time t, th_j is a constant excitation threshold for neuron j, $x_j(t)$ is the activity of neuron j at time t, ap_j is the strength of an action potential (which may be negative for inhibitory units, although none are used in the simulations described here), d_j is the parameter governing the time decay of the potential variable for neuron j, w_{ij} is the strength of the synapse of neuron i on neuron j and r is the refractory period (apart from the time step during which the neuron fires, the output $x_j(t)$ is zero during the refractory period).

The suggestion that the maximum power of computation which could potentially be achieved by a system subject to noise occurs at its transition from ordered to chaotic dynamics has been studied experimentally by a number of authors. Approaches which involve producing systems capable of solving particular computational problems [224] in cellular automata are beset by difficulties

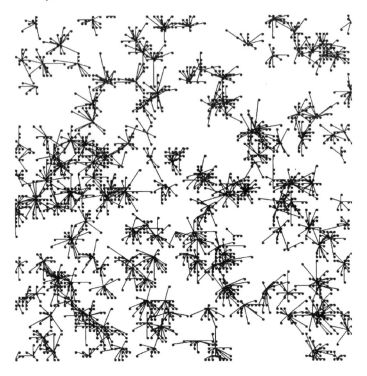

Figure 3.2. Typical connection patterns in a two-dimensional network. Only 3% of neurons are shown for clarity.

in characterizing the dynamics of these discrete space and time systems [174] and by determining whether it is those dynamics themselves, rather than their effect on the algorithms used to search the systems' parameter space, which determines the capabilities of the systems found [198]. An alternative to studying the solution of explicit computational problems is to characterize the implicit computational complexity of systems, that is, to determine the computational power of the simplest machine which can describe the behaviour of the system irrespective of any interpretation of that behaviour [77]. The aim is not, therefore, to find a way of achieving a particular computation, but rather of finding what class of computation the system would be capable of if a suitable interpretation could be found. Application of this method to very simple dynamical systems shows a qualitative difference, between the implicit finite-state computation found in both ordered and chaotic regimes and the computational power of the type of stack-machine required to describe the dynamics of the system at the transition between ordered and chaotic regimes [78].

Recent work on the phenomenon of self-organizing criticality suggests that, in systems similar to the networks described here, maintaining dynamics at the transition between order and chaos might not need particularly fine parameter tuning [23, 21]. I have shown that the networks described in this chapter can similarly be maintained in such a regime over a wide range of parameters. This should be unsurprising given the similarities between the network model and other systems exhibiting self-organizing criticality [64, 22].

3.10   Measuring the intrinsic computation in the network

Recent work by Crutchfield and Young [77, 78] may make practical the seemingly intractable task of objectively measuring the implicit computational power of a complex dynamical system like the neural network.

We usually expect computation to involve discrete symbols and rules, whereas dynamical systems are much more general, covering systems with both continuous and discrete space and time evolution. The symbolic dynamics of a system with continuous space or time variables can be investigated by discretely sampling or averaging those variables to produce a version of the system with completely discrete dynamics. This is the initial stage of Crutchfield and Young's algorithm. Once we have produced a discrete picture of a system's dynamics we then need to abstract a set of rules which describe those discrete dynamics. Rule abstraction can be based upon two separate views of the fundamental processes governing a system's dynamics: deterministic or probabilistic. Rule abstraction from discrete sequences is, perhaps, most usually encountered in the concepts of algorithmic complexity [62]; here the aim is to find the most concise description of a discrete sequence of symbols—the rate of growth of the size of this description with the length of string to be described is an index of the complexity of the string. The most concise descriptions will often be in terms of sets of mappings between symbol sequences in the string and new sets of meta-symbols and of rules governing the allowable transitions between meta-symbols. These rules are deterministic; the outcome of this is that a symbol sequence generated from a chaotic data stream will always have maximal complexity (because no prior sequence of symbols is guaranteed to predict the next symbol deterministically, so the only effective description of the string is the string itself) even though we know the sequence was produced by a mathematically simple process and we know that only a limited set of states can ever be encountered. The induction of purely deterministic rules often fails to capture a useful picture of a dynamical system in symbolic computational terms. Probabilistic symbolic descriptions of dynamics can, however, capture important features of dynamical systems symbolically without being obscured by every detail, detecting, for example, the two distinct orbits in Rössler's bands (see figure 3.3) and the relative probabilities of orbiting within and switching between them. Crutchfield and Young's algorithm produces probabilistic descriptions of this type, although, if presented with data which is described by a simple deterministic process then an appropriate deterministic reconstruction will be produced.

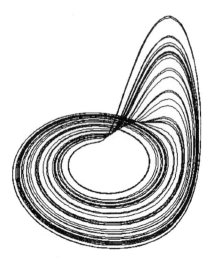

Figure 3.3. Rössler's bands.

3.11   The Crutchfield and Young algorithm

The aim of the algorithm is to produce the minimum complexity symbolic description of a dynamical system relative to a random register Turing machine. The output of the system is the formal machine equivalent (referred to by Crutchfield and Young as an ϵ-machine) of a stochastic grammar. Essentially the algorithm depends on finding the set of time invariances in a system's dynamics which have the simplest (minimum entropy) set of transitions between them. The algorithm produces an ϵ-machine from a system's dynamics according to the following procedure:

Produce a quantized time series from the system's dynamics by dividing its state space into regions and sampling its position in state space in terms of these regions at fixed time intervals.

$$I_G = \sum_{v \in V} p_v \sum_{s \in A} p(s|v) \sum_{v' \in V} p(v'|v; s) \log p(v'|v; s)$$

where V is the set of vertices in which $v \to v'$ is a particular transition whose edge is labelled s, $p(v'|v; s)$ is the probability of that transition and $p(s|v)$ is the probability that s is emitted on leaving v. This is simply the weighted conditional entropy of the graph. It sums the entropies of all individual transitions between nodes $(p(v'|v; s)) \log(p(v'|v; s))$ weighted by the probability of encountering the output node $(p(v))$ times the probability that the symbol labelling an edge from that node will occur at that node $(p(s|v))$.

3.12   Symbolic reconstructions of cyclic and chaotic dynamics

As the machine reconstructed by this algorithm is stochastic the algorithm can produce concise symbolic descriptions of both chaotic and limit-cycle systems. Figure 3.5 shows some ϵ-machine reconstructions from limit cycles produced by an iterated logistic map $(X_{t+1} = rX_t(1 - X_t))$ (see figure 3.4). Figure 3.6 shows ϵ-machine reconstructions from chaotic regions with single, multiple and merging chaotic bands. In these cases the ϵ-machine reflects the different probability densities of value ranges of X in the map and the transitions between those regions.

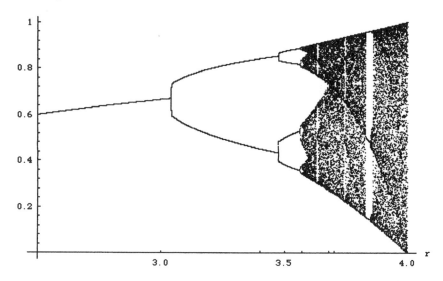

Figure 3.4. Diagram showing the long-time behaviour of the iterated logistic map $X_{t+1} = rX_t(1 - X_t)$.

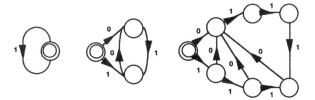

Figure 3.5. ϵ-machine reconstructions of period 1, 2 and 4 limit cycles from the iterated logistic map $X_{t+1} = rX_t(1 - X_t)$ with $r = 2.5$, 3.4 and 3.5 respectively.

Note that all of these machines are finite—they correspond to regular languages—the lowest level of the Chomsky hierarchy. ϵ-machines

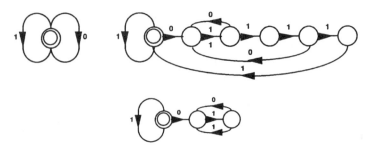

Figure 3.6. ϵ-machine reconstructions of single-band chaos, typical multiple-band chaos and the merger from double to single chaotic bands in the iterated logistic map $X_{t+1} = r X_t (1 - X_t)$ with $r = 4.0$, 3.7 and 3.678 59 respectively.

corresponding to higher-level grammars can, however, be produced by the Crutchfield and Young algorithm. If a system's dynamics is optimally described by a higher-level grammar the algorithm produces truncations of an infinite machine corresponding to that grammar. Regularities in the structure of this machine may allow us to infer the data structures (stacks or tapes) which would allow the machine to be represented in finite terms and the grammar to which the machine corresponds. We can distinguish truncations of infinite machines from finite machines by examining the machines produced from a single data set using successively larger tree depths in the reconstruction algorithm. If the optimal ϵ-machine is finite then this machine will eventually be produced using a particular tree-depth and when reconstructions are performed using deeper trees the same machine will be produced. On the other hand, if the optimal ϵ-machine is infinite then successively larger and larger truncations of the infinite machine will be produced as deeper trees are used in the reconstruction. An example of this can be seen in reconstructions of the ϵ-machine produced from the logistic map at the transition between periodic and chaotic behaviour shown in figure 3.7. In order to make regularities in the growth of this machine apparent the digraph is decorated by collapsing together all deterministic transitions between nodes.

    These examples show that in order to identify whether a system such as a neural network is performing intrinsic computation in the context-free grammar class or above then our criterion must be that the minimal indeterminacy machine size does not reach a limit as reconstruction depth is increased.

3.13   Applying Crutchfield and Young to the neural network

It may appear impractical to apply the Crutchfield and Young algorithm to a network consisting of thousands of neurons (although it is in principle possible) when it can take considerable computational resources to reconstruct ϵ-machines from one-dimensional maps in some parameter ranges. If, however, there is some systematic collective behaviour occurring in the network then this should

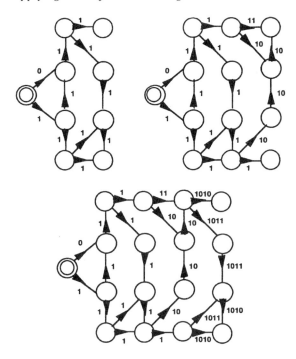

Figure 3.7. ϵ-machine reconstructions of the transition to chaos in the iterated logistic map $X_{t+1} = rX_t(1 - X_t)$ with $r = 3.570$, at reconstruction depths of 5, 9 and 17.

be reflected in the behaviour of single neurons if sampled over a sufficiently long time. I now present results of machine reconstructions from time series of membrane potentials of a single unit from a network showing critical dynamics sampled over 100 000 time steps. As a control the same neuron was sampled under identical driving conditions but disconnected from all other units. First of all I present some data which give a general impression of the behaviour of the network and of a single neuron in isolation.

An inter-spike-interval histogram (as shown in figure 3.8) gives a good conception of an individual unit's behaviour when embedded in the network. This shows the frequencies of intervals between successive firings of the neuron taken from a time series recording of its behaviour.

The same neuron when isolated produces a constant inter-spike interval of 18 as would be expected from its simple evolution equation and constant driving current. It is also obvious that the effect of being embedded in an excitatory network can only reduce a units inter-spike interval. The aim of our computational analysis is to discover what kind of regularities there are in this effect.

Initially I will apply the ϵ-machine reconstruction algorithm to the isolated

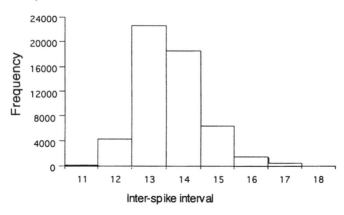

Figure 3.8. Inter-spike interval histogram collected from an individual unit in a network over approximately 50 000 time steps.

neuron. The relative simplicity of its behaviour allows the meaning of the machine reconstructions to be understood easily. At the maximum possible sampling rate we can use for reconstruction (single simulation time-steps) we can choose a potential value at which to code zeros and ones in the initial reconstruction string which discriminates perfectly between the refractory and active phases of the neuron. This discretization value falls at a potential between 0 and 0.4 for the current simulation. The machine reconstructed at this sampling rate and using this discretization value is shown in figure 3.9.

This machine was reconstructed to a sub-tree depth of 19 in order to capture the full 18-step periodicity of the unit's behaviour. The indeterminacy iG is zero. The first two layers of reconstruction establish the initial point in the time-series at which a tree is encountered. Beyond that the paths by which four refractory steps are always followed by 14 active ones can clearly be traced around the machine reconstruction.

Although this reconstruction is easy to relate to the neuron's membrane-potential time-series, we can produce simpler reconstructions. If, instead of discretizing in order to discriminate refractory and active phases of the neuron's behaviour, we discretize so that half of the labels in our initial string are zeros and half are ones by choosing a value near to half of the neuron's firing threshold (in this case about 2.5) then much shallower reconstructions can capture the full dynamics of the neuron as shown in figure 3.10.

Instead of needing to reconstruct to a sub-tree depth of 19 an invariant zero-indeterminacy machine is produced at depths of 10 and above.

In addition to changing the discretization value we can also change our sampling rate in order to further simplify the reconstructed machine. The machine shown in figure 3.11 was produced using a sampling rate four times as fast as that used in the previous examples.

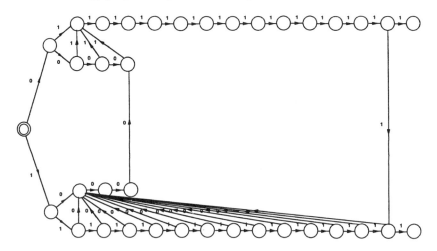

Figure 3.9. Machine reconstructed from the membrane potential time-series of an isolated model neuron. The sampling rate and potential discretization were chosen to highlight the refractory and active phases of the neuron's behaviour. A sub-tree depth of 19 was used in the reconstruction.

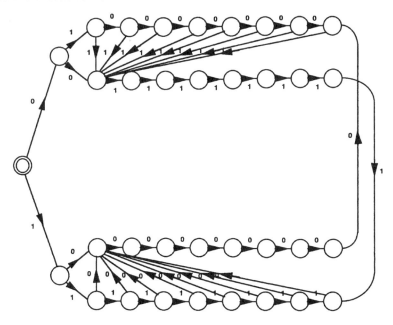

Figure 3.10. Machine reconstructed from the membrane potential time-series of an isolated model neuron. The sampling rate and potential discretization were chosen to minimize the reconstruction depth required to describe the neuron's behaviour. A sub-tree depth of 10 was used in the reconstruction.

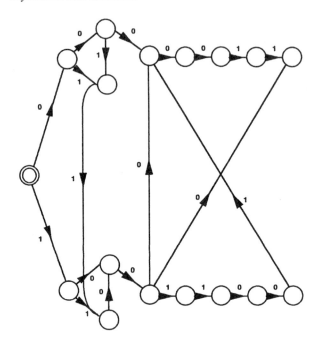

Figure 3.11. Machine reconstructed from the membrane potential time-series of an isolated model neuron. Potential discretization is as before; however data are only sampled every fourth time-step. A sub-tree depth of 8 was used in the reconstruction.

This appears to be the most concise symbolic description of the neuron's behaviour. Higher sampling rates fail to discriminate the refractory period of the neuron (which also represents the strongest non-linearity in its otherwise almost linear behaviour) and hence tend to produce non-zero indeterminacy reconstructions.

We now turn to reconstruction of the behaviour of the same neuron when it is embedded in the network. A reconstruction using exactly the same parameters as used in figure 3.10 is shown in figure 3.12.

The difference between figures 3.11 and 3.12 is striking. The machine reconstructed from the isolated neuron is clearly finite whereas the neuron embedded in the network produces a machine which appears to grow as reconstruction depth increases, which gives no sign of imminent closure and yet which still has zero indeterminacy. Note, however, that reconstruction was only possible to a sub-tree depth of 6 as the number of possible paths through the data becomes extremely large for the embedded neuron and computing deeper reconstructions becomes very time and space intensive.

It is also possible to produce machine reconstruction by producing strings labelled 1 whenever an action potential has just occurred and 0 otherwise. This

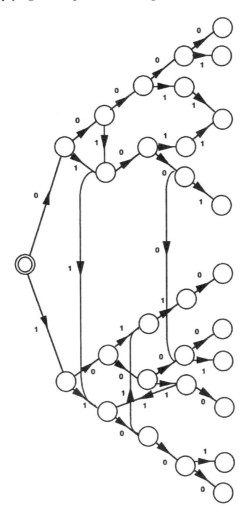

Figure 3.12. Machine reconstructed from the membrane potential time-series of a model neuron embedded in a network. Reconstruction parameters were the same as those used before. A sub-tree depth of 6 was used in the reconstruction.

initial step does not follow Crutchfield and Young's algorithm; the rest of the reconstruction is, however, standard. This method produces far fewer initial paths through the data, so a depth 10 reconstruction of the embedded neuron's behaviour was possible. This is particularly useful since, at a sampling rate of four time-steps, machine closure due to the maximum periodicity of 18 time-steps would be expected at a depth of no more than 8. The depth 10 machine produced is shown in figure 3.13.

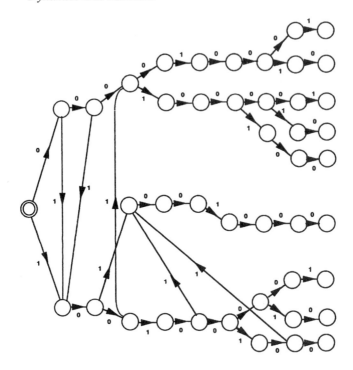

Figure 3.13. Machine reconstructed from an action-potential time-series of a model neuron embedded in a network. A sub-tree depth of 10 was used in the reconstruction.

This reconstruction also appears to be a truncation of an infinite machine; moreover, it shows the lengthening sequences of deterministic transitions which are typical signs of data structures such as stacks or tapes. Its indeterminacy was zero. These findings strongly suggest that the behaviour of this single neuron, when embedded in a network, is optimally described by, and is capable of implementing, computation at the level of context-free grammars or above.

Before accepting this conclusion, however, alternative explanations for this complex behaviour must be investigated. Would the simple presence of multiple inter-spike interval periodicities in the data stream produce such behaviour? This alternative explanation can easily be tested using surrogate data.

Figure 3.14 shows the invariant machine reconstruction from surrogate data in which three periodicities, 4, 5 and 6, were mixed with different probabilities. Each choice of period was independent of previous ones. It can clearly be seen that the reconstructed machine is finite. In fact, it does not differ in form from a period 6 reconstruction; the only influence of the shorter periodicities is to change the transition probabilities in the graph.

It might still be argued that the strong non-linearity of refractoriness combined with multiple independent periodicities might produce complex

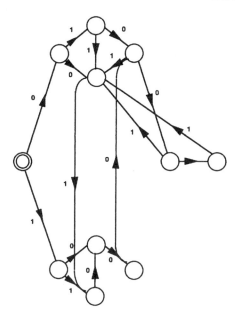

Figure 3.14. Machine reconstructed from surrogate data in which periodicities of 4, 5 and 6 follow one another randomly with different occurrence probabilities of 0.3, 0.4 and 0.3 respectively.

machines such as that reconstructed from the neuron embedded in the network. Figure 3.15, a machine reconstructed from surrogate data with three independent periodicities of 4, 5 and 6 each followed by a fixed two-step refractory period, shows that this is not the case.

The conclusions we can reach from these ϵ-machine reconstructions are that the collective effect of the network on an individual neuron's dynamics is to raise its intrinsic computation to the level of context-free grammars or above. This effect is due to the production of multiple periodicities; moreover, the length of each inter-spike interval must depend on the previous behaviour of the neuron. This dependence continues to outweigh a simple random prediction of all of the neuron's subsequent behaviour in minimizing graph indeterminacy even at large reconstruction depths. Such long-range influences are typical of systems with critical dynamics lending credence to the hypothesis that neural networks in a critical regime might perform computation beyond regular grammars.

3.14 Discussion

We set out to discover what kind of system might share the useful features of both symbolic and artificial neural-network approaches to artificial intelligence. Going beyond the problem of variable binding in neural nets I have argued that

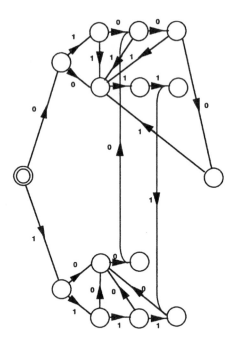

Figure 3.15. Machine reconstructed from surrogate data in which periodicities of 4, 5 and 6 follow one another randomly with different occurrence probabilities of 0.3, 0.4 and 0.3 respectively. Each period is followed by a fixed two-step refractory period.

the usual limitations of artificial neural networks to finite-state computational power must also be transcended. The simulations described have, however, shown that a neural-network model sharing many features with the cerebral cortex exhibits time-dependences in its behaviour which elevates its intrinsic computational power above that of finite-state automata. The key feature of this network which allows such complex behaviour is the spontaneous maintenance of its dynamics at the transition between order and chaos. This produces behaviour which is structured in time; more importantly this structure appears to keep growing over time rather than reaching a finite limit where it is swamped by noise. Although no attempts have been made to understand how to adapt this network to exploit this structure in performing specific computations we can still suggest a possibility and draw some inferences about the style of computation such a system would support.

One way of viewing a critical dynamics is to envision a landscape of attractors which becomes flatter and flatter as criticality is approached. In a near-critical system, which our network must be, these attractors can still influence the transient path which the network follows through its state space although they do not capture it. This is only possible if there is nearly balanced competition between the dynamic processes forming these attractors and other processes which destabilize them. A network in which the probability of connections between neurons depends on their spatial separation must in some sense have some symmetry in its connectivity since the probability of connection between one neuron and another is the same as that between them in the opposite direction. This approximation to symmetry breaks down for individual pairs of neurons which make very few connections with one another but, by the law of large numbers, it holds if we consider the average connectivity between large spatially localized groups of neurons. The approximation is one of connectivity, not weight; however, this does allow a suitable learning process to adapt the average weight between *groups* of neurons to be symmetric. The key point here is that symmetric connection strengths in networks form the basis of point attractor dynamics [343] and moreover the type of adaptive process which produces such connections can be based on biologically plausible activity correlational detecting mechanisms (albeit at a volume rather than individual cell level). Hence competition can exist between large-grain-size attractor dynamics and unstable local dynamics. This hypothetical mechanism for the dynamics of the network described in this chapter is clearly testable and leads to interesting possibilities for learning mechanisms and implications about the extent to which syntax and semantics are separable in such a system.

Acknowledgments

This research was supported by DRA Fort Halstead, UK, contract number 2051/047/RARDE. I would also like to thank the Santa Fe Institute for supporting my visit to their *Seventh Annual Complex Systems Summer School*.

Chapter 4

Information-Theoretic Approaches to Neural Network Learning

Mark D Plumbley
Kings College London, UK
m.plumbley@kcl.ac.uk

4.1   Introduction

It is now nearly 50 years since Shannon first formulated his 'Mathematical theory of communication' [278], where he first introduced *information theory*, which has been used and developed extensively by communications engineers ever since. Communications engineering is concerned with transmitting information from one place to another as efficiently as possible, given certain costs and constraints which are imposed on the communications system which we wish to use. For example, we may have a maximum number of *bits* (binary digits) per second that we can send down a certain binary transmission link, or we may have a radio transmitter with a limit on the maximum power level which we can use. Either of these define constraints which we must work within.

Shannon's information theory gave communication engineering a precise meaning to the idea of *rate of information transmission*. This helped to explain how the properties of a communication channel can limit how fast information can be transmitted through it, and how to *code* signals to make most efficient use of such a channel. One of these results showed that a channel has an innate limit on its information rate, its *capacity*. It is impossible to send information through a channel faster than that channel's *capacity*, no matter how the information is represented or coded.

Other results from information theory deal with *noise* in a communication channel, which may disrupt or degrade a signal. In particular, the presence of noise reduces the information capacity of a channel. However, it is possible to use *error-correcting codes* to reduce the probability of an error in a noisy channel to as close to zero as we like.

Almost as soon as information theory first appeared, psychologists and physiologists were interested in the idea that information theory could help to explain the mechanisms of perception. The visual system of a living creature, for example, could be transmitting information in some form to higher centres of the brain. Treating this visual system as a communications system might help us to understand some of the details behind the function it performs.

Early on, Attneave [18] suggested that visual perception is the construction of an economical description of a scene. Using a guessing game to measure information, he suggested that information in a visual scene is concentrated around the edges and corners of an image, since they are the least predictable from their surroundings. This is consistent with the structural arrangement of simple cells now known to exist in the visual cortex.

More recently, with the resurgence of the field of neural networks, authors such as Linsker [177], Barlow and Földiák [33], Plumbley and Fallside [240], and Atick and Redlich [15] have continued the use of information theory in neural networks, with considerable success. Information theory has proved particularly useful in the development of *unsupervised* learning algorithms. Unlike supervised learning algorithms such as error back-propagation ('BackProp'), unsupervised algorithms have no 'teacher' output to specify what the output of the network should produce. It is therefore more difficult to determine what task or function such a network should learn to perform.

This chapter is organized as follows. Section 4.2 gives a brief introduction to information theory, and introduces Linsker's *Infomax* principle, and information loss. Section 4.3 shows how networks which perform *principal-components analysis* (PCA) can be viewed from this information-theoretic perspective, by making assumptions about the type of input noise. Section 4.4 shows how inhibitory interneurons can be used in this framework, when there is noise on the output of the network. Finally, section 4.5 introduces Becker and Hinton's *I-Max* which extracts depth information from stereograms by maximizing information between neighbouring image patches.

4.2 Information theory

4.2.1 Entropy and information

The two central concepts of information theory are those of *entropy* and *information* [278]. Generally speaking, the entropy of a set of outcomes is the uncertainty of our knowledge about which outcome will actually happen: the less sure we are about the outcome, the higher the entropy. If we know for sure what the outcome will be, the entropy will be zero.

Information is gained by reducing entropy, for example by making an observation of an outcome. Before the observation, our knowledge of the

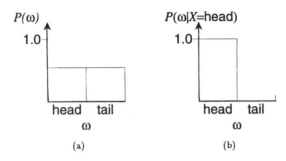

Figure 4.1. Probabilities of coin state $\Omega = \omega$ for (a) before observation and (b) after observation of a 'head'.

outcome is limited, so we have some uncertainty about it. However, after the observation the entropy (uncertainty) is reduced to zero: the difference is the information gained by the observation. A concrete example will be given in a moment, after we introduce the formulas for entropy and information.

Consider an experiment with N possible outcomes i, $1 \le i \le N$ with respective probabilities p_i. The entropy H of this system is defined by the formula

$$H = -\sum_{i=1}^{N} p_i \log p_i \qquad (4.1)$$

with $p \log p$ equal to zero in the limit $p = 0$.

For example, for a fair coin toss, with $N = 2$ and $p_1 = p_2 = \frac{1}{2}$, we have

$$H = -\left(\tfrac{1}{2} \log \tfrac{1}{2} + \tfrac{1}{2} \log \tfrac{1}{2}\right)$$
$$= \log 2.$$

If the logarithm is taken to base 2, this quantity is expressed in 'bits', so a fair coin toss has an entropy of 1 bit.

For any number of outcomes N, the entropy is maximized when all the probabilities are equal to $1/N$. In this case, the entropy is $\log N$. If one of the outcomes has probability 1 with all others having probability 0, then the entropy H in (4.1) is zero: otherwise, H is always positive.

As we mentioned before, the information gained by an observation is the entropy before it, less the entropy after it. As an example, consider our coin toss again, and assume that we observe the outcome to be a 'head'. We denote the state of the coin by the random variable Ω, and write the entropy of the coin toss before any observation by $H(\Omega)$ as in figure 4.1(a). If we denote the observed face by X, we write the *conditional entropy* of the coin after the observation as $H(\Omega|X = \text{'head'})$ as in figure 4.1(b).

The situation if the outcome is a 'tail' is exactly the same. The information

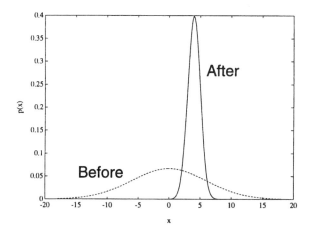

Figure 4.2. Probabilities of a Gaussian distribution before and after a noisy observation. The 'before' distribution has the signal entropy $H(\Omega)$, while the 'after' distribution has the noise entropy $H(\Omega|X)$ for an example observation $X = 4$.

in the observation X about the coin state Ω is then written

$$I(\Omega, X) = H(\Omega) - H(\Omega|X) = \log 2$$

i.e. one bit of information was gained by the observation.

For continuous variables, we cannot use the original discrete-variable formula (4.1), since we now have an infinite number of possible states, which would lead to infinite entropy. Instead we use an alternative form

$$H = -\int_{-\infty}^{\infty} p(x) \log p(x) \, dx \qquad (4.2)$$

which is normally finite, but no longer guaranteed to be positive, and is also dependent on the scaling of variables: scaling by n will add $\log n$ to the entropy.

The information $I(\Omega, X) = H(\Omega) - H(\Omega|X)$ derived from this continuous case *is* scale independent, however, since the scaling will add the same value to both 'before' and 'after' entropies, as in figure 4.2. The entropy $H(\Omega|X)$ represents the noise in the observation X. As an example, for a Gaussian signal of variance $\sigma_S^2 = S$ and noise of variance $\sigma_N^2 = N$, we can calculate that the mean information gained from an observation is

$$I = 0.5 \log(1 + S/N)$$

where S/N is the signal to noise power (variance) ratio. As the noise power N goes to zero, the information gained becomes infinite: so if we could measure a continuous quantity with complete accuracy, we would gain an infinite amount of information. Consideration of noise is therefore very important when determining the information available from a continuous value.

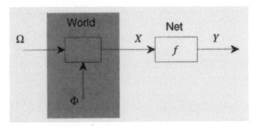

Figure 4.3. The original signal Ω is corrupted by irrelevant noise Φ to give the stimulus X. This is then transformed by the network function f to give the output Y.

4.2.2 Infomax and information loss

If we think of the early parts of a perceptual system such as vision as a system for transmitting information about the environment on to higher centres, it seems reasonable that the more information which is transmitted, the more effective the system will be. Of course, some visual systems are optimized to extract information about very specific stimuli early on: an example would be the 'bug detectors' found in the frog [165]. For higher animals, however, it is more likely that early parts of the visual system should process all input information equally. Linsker therefore suggested his *Infomax* principle: that a perceptual system should attempt to organize itself to maximize the rate of information transmitted (in bits per second) through the system [177].

An alternative view introduced by Plumbley and Fallside [240] is to try to *minimize* the *loss* in information about some original signal Ω as the sensory input is processed by the perceptual system or neural network. Although this approach is in many ways equivalent to Linsker's Infomax principle, it allows a minimax approach to be used in cases when the signal is not Gaussian, for example.

Information loss about Ω across the system which transforms X to Y shown in figure 4.3 is denoted by

$$\Delta I_\Omega(X, Y) = I(X, \Omega) - I(Y, \Omega) \tag{4.3}$$

and has the following properties:

(i) ΔI is positive across any function f, such that $Y = f(X)$;
(ii) ΔI is positive across any additive noise Φ, such that $Y = X + \Phi$ as in figure 4.4(a);
(iii) ΔI is additive across a chain of networks as in figure 4.4(b).

So, to minimize the information loss across a series of networks, the information loss across each network should be minimized. Once information is lost, it cannot be regained.

For an example of this approach, in the next section we examine networks which extract the *principal components* of the inputs. These are often used as a

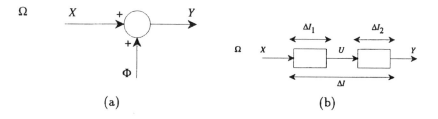

Figure 4.4. Information loss is (a) positive across additive noise and (b) additive in series.

first processing stage, to simplify a pattern recognition problem by reducing the number of dimensions. We shall see what happens when we try to optimize the information transmitted across this type of network.

4.3 Principal-components analysis

Principal-components analysis (PCA) is widely used for dimension reduction in data analysis and pre-processing, and is used under a variety of names such as the (discrete) Karhunen Loève transform (KLT), factor analysis, or the Hotelling transform in image processing. Its primary use is to provide a reduction in the number of parameters used to represent a quantity, while minimizing the error introduced by doing so. In the case of PCA, a purely linear transform is used to reduce the dimensionality of the data, while minimizing the mean squared reconstruction error. This is the error which we get if we transform the output y back into the input domain to try to reconstruct the input x so that the error is minimized.

Linsker's principal of maximum information preservation, Infomax, can be applied to a number of different forms of neural network. The analysis, however, is much simpler when we are dealing with simple networks, such as binary or linear systems. It is instructive to look at the linear case of PCA in some detail, since much effort in other fields has been directed at linear systems. We should not be too surprised to find a neural network system which can perform KLT and PCA.

From one point of view, these conventional data processing methods let us know what to expect from a linear unsupervised neural network. However, the information-theoretic approach to the neural network system can help us with conventional data processing methods. In particular, we shall find that a dilemma in the use of PCA, known as the *scaling problem*, can be clarified with the help of information theory.

4.3.1   The linear neuron

Arguably the simplest form of unsupervised neural network is an N-input, single-output linear neuron as shown in figure 4.5.

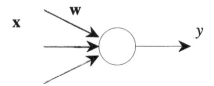

Figure 4.5. The Oja neuron.

Its output response y is simply the sum of the inputs x_i multiplied by their respective weights w_i, i.e.

$$y = \sum_{i=1}^{N} w_i x_i \tag{4.4}$$

or, in vector notation,

$$y = w^T \cdot x \tag{4.5}$$

where w and x are column vectors. The output y is thus the dot product of the input x with the weight vector w. If w is a unit vector, y is the component of x in the direction of w as shown in figure 4.6.

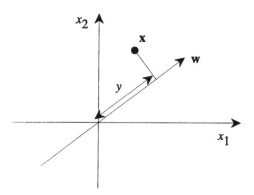

Figure 4.6. Output y as a component of x, with unit weight vector.

We thus have a simple neuron which finds the component of the input x in a particular direction. We would now like to have a neural network learning rule for this system, which will modify the weight vector depending on the inputs which are presented to the neuron.

4.3.2 The Oja principal component finder

Many simple learning rules are based on Hebb's principle [135], which states that the effectiveness of a connection between two neurons should be increased when they are both active at the same time. A very simple form of Hebbian learning rule for the connections in this neuron would be to update each weight by the product of the activations of the units at either end of the weight. For the single linear neuron, this would result in a learning algorithm of the form

$$\Delta w_i = \eta x_i y \tag{4.6}$$

or in vector notation

$$\Delta w = \eta x y \tag{4.7}$$

where η is a learning rate parameter. Unfortunately, this learning algorithm alone would cause any weight to increase without bound, so some modification has to be used to prevent the weights from becoming too large.

One possible solution is to limit the absolute values that each weight w_i can take, while another is to renormalize the weight vector w to have unit length after each update [220]. An alternative is to use a weight decay term which causes the weight vector to tend to have unit length as the algorithm progresses, without explicitly normalizing it. To see how this works, consider the following weight update algorithm, due to Oja [220]:

$$\begin{aligned} \Delta w &= \eta(x y - w y^2) \\ &= \eta(x \cdot x^T \cdot w - w \cdot (w^T \cdot x \cdot x^T \cdot w)). \end{aligned} \tag{4.8}$$

When the weight vector is small, the update algorithm is dominated by the first term on the right-hand side, which causes the weight to increase as for the unmodified Hebbian algorithm. However, as the weight vector increases, the second term (the 'weight decay' term) on the right-hand side becomes more significant and this tends to keep the weight vector from becoming too large.

To find the convergence conditions of the Oja algorithm, let us consider the average weight update over some number of input presentations. We shall assume the input vectors x have zero mean and we shall also assume that the weight update factor is so small that the weight itself can be regarded as approximately constant over this number of presentations. Thus the mean update is given by

$$\begin{aligned} \overline{\Delta w} &= \eta \left(\overline{x \cdot x^T \cdot w} - \overline{w \cdot w^T \cdot x \cdot x^T \cdot w} \right) \\ &\approx \eta \left(\overline{x \cdot x^T} \cdot w - w \cdot \left(w^T \cdot \overline{x \cdot x^T} \cdot w \right) \right) \\ &\approx \eta \left(\mathbf{C}_x w - w \lambda \right) \end{aligned} \tag{4.9}$$

where $\lambda = w^T \mathbf{C}_x w$ and $\mathbf{C}_x = E(x \cdot x^T)$ is the covariance matrix of the input data x.

When the algorithm has converged, the average value of Δw will be zero, so we have

$$\mathbf{C}_x w = w\lambda \tag{4.10}$$

i.e. the weight vector w is an eigenvector of the input covariance matrix $\mathbf{C}_x$. A perturbation analysis confirms that the only stable solution is for w to be the *principal* eigenvector of $\mathbf{C}_x$. To find the eventual length of w we simply substitute (4.10) into the expression for λ and we find that

$$\lambda = w^T \cdot (\mathbf{C}_x w) = w^T \cdot (w\lambda) \tag{4.11}$$

i.e. provided λ is non-zero, $w^T \cdot w = 1$ so the final weight vector has unit length.

We have therefore seen that as the Oja algorithm progresses, the weight vector will converge to the normalized principal eigenvector of the input covariance matrix (or its negative). The component of the input which is extracted by this neuron, to be transmitted through its output y, is called the *principal component* of the input and is the component with largest variance for any unit length weight vector.

4.3.3   Reconstruction error

For our single-output system, suppose we wish to find the best estimate $\hat{x}$ of the input x from the single output $y = w^T \cdot x$. We form our reconstruction using the vector u as follows:

$$\hat{x} = uy \tag{4.12}$$

where u is to be adjusted to minimize the mean squared error

$$\epsilon = E\left[|x - \hat{x}|^2\right] = E\left[|(\mathsf{I} - u \cdot w^T)x|^2\right] \tag{4.13}$$

where I is the identity matrix. If we minimize ϵ with respect to u for a given weight vector w, we get a minimum for ϵ at

$$u_w = \arg\min_u(\epsilon) = \frac{\mathbf{C}_x w}{w^T \mathbf{C}_x w} \tag{4.14}$$

where $\mathbf{C}_x = E[x \cdot x^T]$ as before (assuming that x has zero mean). Our best estimate of x is then given by

$$\hat{x}_w = u_w y = \frac{\mathbf{C}_x w \cdot w^T}{w^T \mathbf{C}_x w} x = \mathbf{Q}x \tag{4.15}$$

where the matrix

$$\mathbf{Q} = \frac{\mathbf{C}_x w \cdot w^T}{w^T \mathbf{C}_x w} \tag{4.16}$$

is a *projection* operator, a matrix operator which has the property that $Q^2 = Q$. This means that the best estimate of the reconstruction vector $\hat{x}_w$ from the output is $\hat{x}_w$ itself. Once this is established, it is possible to minimize ϵ with respect to the original weight vector w. Provided the input covariance matrix C_x is positive definite, this minimum occurs when the weight vector is the principal eigenvector of C_x. Thus PCA minimizes the mean squared reconstruction error.

4.3.4 The scaling problem

Users of PCA are sometimes presented with a problem known as the *scaling problem*. The result of PCA, and related transforms such as the Karhunen Loève transform (KLT), is dependent on the scaling of the individual input components x_i. When all of the input components come from a related source, such as light level receptors in an image processing system, then it is obvious that all the inputs should have the same scaling. However, when different inputs represent unrelated quantities, then the relative scaling which each input should be given is not so apparent. As an extreme example of this problem, consider two uncorrelated inputs which initially have equal variance. Whichever input has the largest scaling will become the principal component. While this extreme situation is unusual, the scaling problem does cause PCA to produce scaling-dependent results, which is rather unsatisfactory.

Typically, this dilemma is solved by scaling each input to have the same variance as each other [331]. However, there is also a related problem which arises when multiple readings of the same quantity are available. These readings can either be averaged to form a single reading, or they can be used individually as separate inputs. If same-variance scaling is used, these two options again produce inconsistent results.

Thus although PCA is used in many problem areas, these scaling problems may lead us not to trust it to give us a consistent result in an unsupervised learning system.

4.3.5 Information maximization

We have seen that the Oja neuron learns to perform a principal-components analysis of its input, but that principal-components analysis itself suffers from an inconsistency problem when the scaling of the input components is not well defined. In order to gain some insight to this problem, Linsker applied his *Infomax* principle [177] to this situation.

Consider a system with input X and output Y. Linsker's Infomax principle states that a network should adjust itself so that the information $I(X, Y)$ transmitted to its output Y about its input X should be maximized. This is equivalent to the information in the input X about the output Y, since $I(X, Y)$ is symmetric in X and Y.

However, if Y is a noiseless function of X, as is the case for our linear neuron

$$y = w^T \cdot x$$

then there will be an *infinite* amount of information in the output Y about X, because Y represents X infinitely accurately. In order to proceed, we must assume that the input contains some noise ϕ which prevents any of the input from being measured too accurately.

Consider the case where the input signal x is a zero-mean Gaussian with covariance matrix $\mathbf{C}_x$ and the noise ϕ is also a zero-mean Gaussian, with covariance matrix $\mathbf{C}_\phi = \sigma^2 \mathbf{I}$, so that the noise on each input component is uncorrelated with equal variance. The output of the neuron is then the weighted sum

$$y = w^T \cdot (x + \phi) = w^T \cdot x + w^T \cdot \phi. \tag{4.17}$$

Writing down the information in the output y about the input signal x, we get

$$I(Y; X) = \tfrac{1}{2} \log \frac{S + N}{N} \tag{4.18}$$

where

$$S = E\left(|w^T \cdot x|^2\right) = w^T \mathbf{C}_x w$$

and

$$N = E\left(|w^T \cdot \phi|^2\right) = \sigma^2 |w|^2.$$

Since (4.18) is monotonically increasing in S/N, $I(X, Y)$ is maximized when w is the principal eigenvector of $\mathbf{C}_x$, i.e. it is the principal component of the input.

This is the same condition for minimizing the mean squared reconstruction error considered above, but now we have an explicit condition on the noise on the input. The condition is that the noise on each input should be uncorrelated and each input should have the same noise variance.

The scaling problem of principal-components analysis is now changed to one of guessing the noise on each of the inputs and scaling them so that this noise is approximately equal. The standard approach of scaling all inputs so that their signal variance is equal is therefore equivalent to assuming that the signal to noise ratio of all inputs is equal [236].

Of course, the assumptions that the input signal and noise are Gaussian and zero mean are very strong, but can be relaxed somewhat if *information loss* is considered rather than *transmitted information*. However, the result in each case is the same: the Oja algorithm, which finds the principal component of the input, maximizes information capacity on condition that the noise on the input components is equal.

4.3.6   Multi-dimensional PCA

There are a number of algorithms which extend Oja's algorithm to more than one output neuron. For these we need an output vector y and a weight matrix $\mathbf{W}$, such that

$$y = \mathbf{W}^T x. \tag{4.19}$$

If the Oja algorithm was used for each output neuron with no modification, each would find the same principal component of the input data. Some mechanism must be used to force the outputs to learn something different from each other.

One possibility is to use a lateral inhibition network between the output neurons, which forces their outputs to be decorrelated [109]. An alternative is to modify the weight decay term of the Oja algorithm: this approach is used by Williams' symmetric error correction (SEC) algorithm [332], Oja and Karhunen's M-output PCA algorithm [221] and Sanger's generalized Hebbian algorithm (GHA) [267].

In fact, these algorithms have much in common. Although the weight vectors themselves have different algorithms, the subspace defined by the orthogonal projection

$$\mathbf{P} = \mathbf{W}(\mathbf{W}^T\mathbf{W})^{-1}\mathbf{W}^T$$

which is the subspace spanned by the weight vectors to each of the outputs, moves in exactly the same way for each of these algorithms. Since this subspace, rather than the weight vector itself, determines the change in information transmitted through the network, these three algorithms tend to increase the transmitted information in exactly the same way. All three lead to a set of weight vectors which spans the same space as the largest principal eigenvectors of the input covariance matrix, which is sufficient to maximize the transmitted information (under the equal-noise conditions which we outlined above) [239].

The three algorithms differ only in the behaviour of the weight vectors themselves. In particular, the SEC algorithm [332] leads to weight vectors which are orthogonal and of unit length, but which have no particular relationship to the eigenvectors of the input covariance matrix. Oja and Karhunen's algorithm [221] uses a Gramm–Schmidt orthogonalization (GSO) approach to find the principal components themselves, in order. Sanger's algorithm [267] uses GSO in a slightly different way, but also finds the principal components in order.

4.3.7   Uncorrelated outputs

We have seen that linear neurons, with a modified form of Hebbian learning algorithm, can learn to find the principal component or principal subspace of their input data. We have also seen that this principal subspace maximizes the information capacity of the system, under the condition that the input components have uncorrelated, independent, equal-variance Gaussian noise on all of the components.

When we perform principal-components analysis in practice, we also tend to use a set of outputs which are decorrelated, or at the very least not highly correlated with each other. The algorithms considered here also do this, but this does not seem to be required to maximize information capacity. We should only have to find the principal subspace of the input data: correlation between the output components should be irrelevant.

The puzzle here arises because we have neglected noise which may occur *after* the network which we are currently considering. This noise may be due to added noise or calculation errors in stages following the PCA network. In the next section, we shall see that decorrelated outputs tend to be better protected against later noise than outputs which are highly correlated [33]. It may be that real perceptual systems are able to take account of both noise sources at the same time.

4.4   Lateral inhibition and anti-Hebbian learning

In the networks we have considered so far, we have had feedforward connections from one layer to another. In this section, we shall consider networks with *lateral* connections, i.e. connections between units within the same layer.

Soon after Attneave's suggestion [18] that a visual system should create an economical description of a scene, Barlow [32] argued that *lateral inhibition* could be a possible mechanism to achieve this economical description. This would involve inhibitory connections between neurons within a layer. Any signal which is common to many neurons would be reduced by this lateral inhibition effect. Uniform areas of the visual field would produce little output, while edges would produce a significant output from this layer. This lateral inhibition would reduce the amount of *redundant* information, i.e. information contained in more than one signal, and hence produce a more economical representation.

Neural network algorithms have been suggested which can learn to perform this redundancy reduction. These are sometimes called *anti*-Hebbian, since they cause the *inhibition* to increase if the neurons at either end are active.

4.4.1   Decorrelating algorithms

Barlow and Földiák [33] suggested a network with linear recurrent lateral inhibitory connections as shown in figure 4.7(a) with an anti-Hebbian local learning algorithm. This network is designed to *decorrelate* its outputs, i.e. to produce network outputs which are all uncorrelated with each other. Removal of correlations like this is sometimes known as *whitening*.

In vector notation, we have an M-dimensional input vector x, an M-dimensional output vector y and an $M \times M$ lateral connection matrix $\mathbf{V}$. For a fixed input, the lateral connections cause the output values to evolve according

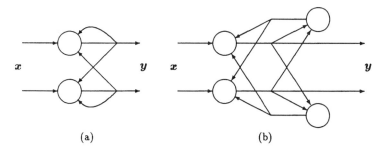

(a)   (b)

Figure 4.7. Linear decorrelating networks ($M = 2$).

to the expression

$$(y_i)_{t+1} = x_i - \sum_j v_{ij}(y_j)_t \qquad \text{i.e.} \qquad y_{t+1} = x - Vy_t \qquad (4.20)$$

at time step t, which settles to an equilibrium when $y = x - Vy$, which we can write as

$$y = (I_M + V)^{-1}x \qquad (4.21)$$

provided $(I_M + V)$ is positive definite. We assume that this settling happens virtually instantaneously. The matrix V is assumed to be symmetrical so that the inhibition from unit i to unit j is the same as the inhibition from j to i and for the moment we assume that there are no connections from a unit back to itself, so the diagonal entries of V are zero.

Barlow and Földiák suggested that for each input x, the weights v_{ij} between different units should be altered by a small change

$$\Delta v_{ij} = \eta y_i y_j \qquad i \neq j \qquad (4.22)$$

where η is a small update factor. In vector notation this is

$$\Delta V = \eta \, \text{offdiag}(y \cdot y^T) \qquad (4.23)$$

since the diagonal entries of V remain fixed at zero. This algorithm converges when $E(y_i y_j) = 0$ for all $i \neq j$, and thus causes the outputs to become decorrelated.

Atick and Redlich [17] considered a similar network, but with an integrating output $dy/dt = x - Vy$ leading to $y = V^{-1}x$ when it has settled. They show that a similar algorithm for the lateral inhibitory connections between different output units leads to decorrelated outputs, while reducing a information-theoretic redundancy measure.

4.4.2 Optimizing information with lateral inhibition

The principal-components algorithms discussed above were concerned with optimizing information given noise on the input to the network. These decorrelating algorithms instead attempt to optimize information in the case of noise on the *output* of the network.

To analyse this, we follow an argument by Shannon [279]. Consider a network with input represented by the random variable X, output Y, with added noise Φ giving a final information-bearing output $\Psi = Y + \Phi$.

For small output noise, we can express the transmitted information as

$$I(\Psi, X) = \frac{1}{2 \log \det \mathbf{C}_Y} - \frac{1}{2 \log \det \mathbf{C}_\Phi} \qquad (4.24)$$

and the power cost as

$$S_T = \text{Tr}(\mathbf{C}_Y). \qquad (4.25)$$

We wish to maximize $I(\Psi, X)$ for fixed S_T, so we use the Lagrange multiplier technique and instead attempt to maximize the function

$$J = I(\Psi, X) - \frac{1}{2\lambda S_T}. \qquad (4.26)$$

This leads to the condition [237]

$$\mathbf{C}_Y = (1/\lambda)\mathbf{I}_M \qquad (4.27)$$

so, not only should the outputs be decorrelated, but they should all have the same variance, $E(y_i^2) = 1/\lambda$.

The decorrelating algorithms outlined above will be sufficient if we know that the signal we are dealing with is not dependent on position. This might be the case for a image on a regular grid, for example, provided images could appear anywhere on this grid. If this is not the case, we may need a slightly different algorithm.

The Barlow and Földiák algorithm can be modified to achieve this, if self-inhibitory connections from each unit back to itself are allowed [237]. The algorithm becomes

$$\Delta v_{ij} = \eta y_i y_j - (1/\lambda)\delta_{ij} \qquad \text{i.e.} \qquad \Delta\mathbf{V} = \eta(\mathbf{y} \cdot \mathbf{y}^T - (1/\lambda)\mathbf{I}_M) \qquad (4.28)$$

which monotonically increases J as it progresses.

This is perhaps a little awkward, since the self-inhibitory connections have a different update algorithm to the normal lateral inhibitory connections. As an alternative, a linear network with inhibitory interneurons as shown in 4.7(b) can be used. After an initial transient, this network settles to

$$\mathbf{y} = \mathbf{x} - \mathbf{V}z \qquad \text{and} \qquad z = \mathbf{V}^T\mathbf{y} \qquad (4.29)$$

i.e.

$$\mathbf{y} = (\mathbf{I} + \mathbf{V}\mathbf{V}^T)^{-1}\mathbf{x} \qquad (4.30)$$

where v_{ij} is now the weight of the excitatory (positive) connection from y_i to z_j and also the weight of the inhibitory (negative) connection back from z_j to y_i.

Suppose that the weights in this network are updated according to the algorithm

$$\Delta v_{ij} = \eta(y_i z_j - (1/\lambda)v_{ij}) \qquad (4.31)$$

which is a Hebbian (or anti-Hebbian) algorithm with weight decay and is

$$\Delta \mathbf{V} = \eta(\mathbf{C}_Y - (1/\lambda)\mathbf{I}_M)\mathbf{V} \qquad (4.32)$$

in vector notation. Then the algorithm will converge when $\mathbf{C}_Y = (1/\lambda)\mathbf{I}_M$, which is precisely what we need to maximize J. In fact, this algorithm will also monotonically increase J as it progresses.

This network suggests that inhibitory interneurons, which are found in many places in sensory systems, may be performing some sort of decorrelation task. Not only does the condition of decorrelated equal variance output optimize information transmission for a given power cost, but it can be achieved by various biologically plausible Hebb-like algorithms.

4.4.3 Introducing more realistic constraints

A real visual system such as the retina has to deal with noise on both the incoming signal (due to photon shot noise) and noise on the output of the network (due to, e.g. spiking shot noise). Under certain simplifying conditions (e.g. the output noise dominates) it is possible to develop networks with simple local algorithms which can optimize information with both noise sources [238].

In more realistic situations it is more difficult to develop learning algorithms, but information theory can still be applied with very interesting results. For example, Atick and Redlich [16] derived a theoretical prediction for the spatial characteristics of retinal filters as a function of background light levels. This theoretical prediction shows a remarkable agreement with human psychophysical contrast sensitivity measurements.

More recently, Dong and Atick [91] have extended this work to the temporal domain, including the effect of nonlinearities in the visual pathway. They suggest that the lateral geniculate nucleus (LGN), until recently considered to be simply a relay station between the retina and the visual cortex, in fact helps to optimize the representation of visual information in the temporal domain. Since the neurons in the LGN can only have a positive firing rate, some cells represent the positive (nonlagged) parts of the signal, while others represent the negative (lagged) parts.

4.5   I-Max: maximizing mutual information between output units

In a slightly different use of information theory to those outlined above, Becker and Hinton [37] suggested that the information *between* output units could be

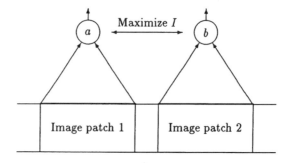

Figure 4.8. Two units with separate inputs that maximize their mutual information.

used as the objective function for an unsupervised learning technique as shown in figure 4.8.

In a visual system, this scheme would attempt to extract higher-order features of the visual scene which are coherent over space (or time). For example, if two networks each produce a single output from two separate but neighbouring *patches* of a retina, the objective of their algorithm is to maximize the mutual information $I(Y_1, Y_2)$ between these two outputs. A steepest-ascent procedure can be used to find the maximum of this mutual information function, both for binary and real-valued units.

One application of this principle is the extraction of depth from random-dot stereograms. Nearby patches in an image usually view objects of a similar depth, so if the mutual information between neighbouring patches is to be maximized, the outputs from both output units y_1 and y_2 should correspond to the information which is common between the patches, rather than that which is different. In other words the outputs should both learn to extract the common depth information rather than any other property of the random dot patterns.

For binary-valued units, with each unit similar to that used by the *G*-max scheme described above, the mutual information $I(Y_1, Y_2)$ between the two output units is

$$I(Y_1, Y_2) = H(Y_1) + H(Y_2) - H(Y_2, Y_2) \tag{4.33}$$

so if the probability distributions $P(y_1)$, $P(y_2)$ and $P(y_1, y_2)$ are measured, this mutual information can be calculated directly. Of course, it is sufficient to measure $P(y_1, y_2)$ only, since

$$P(y_1) = \sum_{y_2=0}^{1} P(y_1, y_2)$$

and similarly for $P(y_2)$. The derivative of (4.33) can be taken with respect to the weights in the network for each different input pattern, so enabling the steepest-ascent procedure to be used.

For real-valued outputs it would be impossible to measure the entire probability distribution $P(Y_1, Y_2)$, so instead it is assumed that the two outputs have a Gaussian probability distribution, and that one of the outputs is a noisy version of the other, with independent additive Gaussian noise. In this case, the information $I(Y_1, Y_2)$ between the two outputs can be calculated from the variances of one of the outputs (the 'signal') and the variance of the difference between the outputs (the 'noise') as

$$I(Y_1, Y_2) = \tfrac{1}{2} \log \frac{\sigma_{Y_1}^2}{\sigma_{Y_1 - Y_2}^2} \qquad (4.34)$$

where $\sigma_{Y_1}^2$ is the variance of the output of the first unit and $\sigma_{Y_1 - Y_2}^2$ is the variance of the difference between the two outputs.

If we accumulate the mean and variance of both Y_1 and $Y_1 - Y_2$, it is possible to find the derivative of (4.34) for each input pattern, with respect to each weight value. Thus the weights in the network can be updated in a steepest-ascent procedure to maximize $I(Y_1, Y_2)$, or at least the approximation to $I(Y_1, Y_2)$ given by (4.34).

Becker and Hinton found that unsupervised networks using this principle could learn to extract depth information from random-dot stereograms with either binary- or continuous-valued shifts, as appropriate for the type of output used, although in some cases it helped to force the units to share weight values, further enforcing the idea that the units should calculate the same function of the input. They generalized their scheme to allow networks with hidden layers, and also to allow multiple output units, with each unit maximizing the mutual information between itself and a value predicted from its neighbouring units. This latter scheme allowed the system to discover an interpolation for curved surfaces.

4.6   Conclusions

In this chapter we have introduced some of the ways that information theory is used with neural networks. We have seen that noise, which is not often important in many neural networks, is of central importance when dealing with information theory. Noise limits the information which can be transmitted by a network. If the input has spherical Gaussian noise, information is optimized if we use a network which extracts the principal components of the input suppressing low-amplitude input components. If noise is on the output, a decorrelating (whitening) network is needed, boosting low-amplitude input components. In more realistic situations, where noise is present on both input and output, a trade-off between these two extremes emerges.

As an illustration of other possibilities, we have also seen how common information can be extracted from network outputs using the 'I-max' approach. Although the depth is only a relatively small amount of the total information in a random-dot stereogram, maximizing the information between the outputs of networks connected to neighbouring patches extracts the depth because no other information is common to both patches.

Although we have concentrated on unsupervised learning in this chapter, information theory can also give us insight into supervised learning. For example, the graded outputs for a multi-layer perceptron contain more information about the likely output than the identity of the 'best' output alone. If the output from this supervised network is to be used in some further processing stage, we could keep more information by, for example, keeping some m best outputs instead of just the single best output [236].

Information theory has proved to be a useful tool in the neural networks armoury. In particular, it has given us a 'driving force' for unsupervised neural network learning algorithms. The arguments that biological information processing systems are optimizing information are backed up by good results from early vision. These suggest that in the future, information theory could offer us a way to investigate how perception works *above* the level of individual neurons, by considering what happens to the *information*, not just to individual signals.

PART 2

COGNITIVE MODELLING

From the very beginning of neural computing, neural networks were used to model aspects of human cognition. It was only later that their usage spilled over into the industrial and commercial sectors. Human reasoning and language are two areas where so far no computational system has reached human performance. Neural networks provide a radically different approach to that of the symbolic computational systems used in the past to model cognitive tasks. This part of the book outlines some of the various approaches to attempting what is no doubt the most complex task facing computer science, that of modelling the human mind.

Chapter 5

Exploring Different Approaches towards Everyday Commonsense Reasoning

Ron Sun
University of Alabama, USA
rsun@cs.ua.edu

5.1 Introduction

This chapter attempts to explore high-level connectionist models for everyday commonsense reasoning from a broad perspective, investigating their connections to existing paradigms. Firstly this chapter examines two other prominent paradigms: rule-based reasoning and case-based reasoning. high-level connectionist models, especially the CONSYDERR architecture, are studied in light of these paradigms. Although the majority of neural network (connectionist) research is concentrated on low-level cognitive processes, such as vision, pattern recognition, and associative memory, connectionist models are also believed to be suitable for modelling high-level cognitive processes, such as commonsense reasoning, everyday planning, and natural language processing. It is important for connectionist models to deal with such high-level cognitive processes for the following two reasons:

(i)   A paradigm has to be able to address these problems in order to qualify as a model of intelligence and cognition.
(ii) The symbolic paradigm has been unable to deal with some fundamental problems in modelling these aspects of intelligence, and thus there is a need for exploring alternatives in order to make some progress.

High-level connectionist models aim at understanding high-level cognitive processes by providing a detailed model cast in network terms. Through the interaction and the propagation of activations among nodes in a network, some high-level cognitive functions can be accomplished by the network as a whole;

such models are often more detailed and more realistic when compared with symbolic models.

These connectionist models usually involve some kind of symbolic processing, albeit implemented in a network fashion. In many ways, high-level connectionist models are stepping into the traditional territory of 'symbolic' AI and blurring the boundary. In order to better understand high-level connectionist models, we need an examination of the history of ideas associated with different paradigms of artificial intelligence so that we can properly place such models into our mental space of ideas, approaches, and paradigms, and see their relative strengths and future directions. We need a comparative study in a certain broad domain.

Notably reasoning (including approximate reasoning) is a fundamental aspect of human cognition: it is involved in all kinds of cognitive processes, ranging from language understanding to decision making. Therefore it is of uttermost importance to achieve a better understanding of this fundamental aspect of cognition. Among various forms of reasoning, everyday commonsense reasoning represents the most basic and prevailing reasoning activity, for it is the most frequently occurring and the most fundamental, relied on in everyday coping with the world and serving as a basis for all other reasoning processes [92]. Therefore it is well justified for us to concentrate on commonsense reasoning (which by itself is an almost all-embracing concept), when examining high-level connectionist models. It is somewhat structured yet flexible (e.g., it can be approximate), and it is usually reliable but sometimes fallible [82]. Commonsense reasoning is also one of the main problems in artificial intelligence. It has been extremely difficult for AI programs to capture commonsense knowledge and reasoning in all its power and flexibility.

Even the very concept, commonsense reasoning, is difficult to characterize: we cannot define what commonsense reasoning is, just as it is hard to define what intelligence is, or what knowledge is. Roughly speaking, however, commonsense reasoning can be taken (at least for the kind of commonsense reasoning explored in this work) as referring to informal kinds of reasoning in everyday life regarding mundane issues, where no exhaustive search is performed and speed is often more important than accuracy.

We will therefore examine different AI paradigms in relation to commonsense reasoning, and attempt to answer the following questions:

- How good is each of them at dealing with reasoning? Do they have the potential for generating robust commonsense reasoning?
- How are these different approaches related to each other? Particularly, how are these approaches related to high-level connectionist models?
- Can we synthesize some of these diverse ideas, each of which excels at certain aspects, so that better, more integrated models for commonsense reasoning can be formed?
- Can high-level connectionist models serve as a unifying framework?

The following sections will look into three different approaches, rules and logics, case-based reasoning, and connectionism, based on certain criteria we hypothesize. We will see that from rule-based reasoning to case-based reasoning and further to connectionist models, a Hegelian cycle of *thesis/antithesis/synthesis* is fulfilled.

5.2   Rule-based reasoning

In this section, we will first discuss rules in general, and then we will discuss the formal logic accounts of reasoning as a special case of the rule-based approach. We will look into some logics for dealing with non-monotonicity in reasoning. We will also discuss fuzzy logic. Finally, we will look into probabilistic views of rule-based reasoning.

5.2.1   Rules in general

Among the paradigms used in the existing accounts of commonsense reasoning, the rule-based paradigm is by far the most prominent.   In this paradigm, generally, some generic, common syntactic forms that consist of *antecedents* and *consequents* (or conditions and conclusions) are used, as a basic representational means [161].

However, this type of account is not free from problems, the most acute of which is the *brittleness* (or rigidity) problem. There are various attempts toward amending it, such as non-monotonic logics, circumscription, and fuzzy logic.

There is a long history of believing that rules can be used to explain human behaviour. This belief lies at the core of the rationalism that has been underlying Western culture [92]. Human cognition is believed in Plato's work to be from rational verbal inferences from facts. *Syllogism, modus ponens*, and *modus tollens* are believed to be the essential forms of such reasoning, according to the early rationalist philosophers, such as Aristotle. These forms were fully elaborated and pushed as far as they can go logically throughout the medieval time, and again later on emphasized by Descartes and other rationalist philosophers in their philosophies of mind. Kant, in *A Critique of Pure Reason* [156], speaks of cognition as pure reason that works on the constructed world built through sensory *lebenswelt* (life-world) with innate frameworks and rules ('synthetic *a priori'*) that determine ubiquitously correct forms of reasoning.   The modern day version of this rationalistic approach in AI is the *physical symbol hypothesis* [212], which spawned (and was used to justify) enormous research effort in symbolic AI. This approach typically uses discrete symbols as basic representational primitives, and performs symbol manipulation for reasoning (mostly based on rules), and typically does it (based on rules) in a sequential, analytic manner, without many complex interactions or any substrate constraints [92] in representation and reasoning. There are certain areas in which symbolic AI achieved certain successes, the most notable of which is rule-based expert systems.

Since rule-based reasoning is the most popular paradigm of symbolic AI so far, there are numerous variants and implementations for symbolic rule-based reasoning, for example, the OPS family of languages [161], the ACT family of production system models [12], and the whole area of expert systems [57].

But a severe problem has been plaguing all of these existing systems in this paradigm, and symbolic AI in general: that is, the *brittleness* of these systems, despite the differences and variations among them. Though different authors have ascribed somewhat different meanings to the word, basically 'brittleness' suggests being easily breakable: the slightest deviation in inputs from what is exactly known by a system can cause a complete breakdown of the system. In general, such systems can only work within a predefined narrow domain with precisely pre-specified situations. Specifically, it can be qualified as the inability of a system to deal with (in a systematic way and within a unified framework) some important aspects in reasoning, including the following [299]:

- Partial, uncertain, or fuzzy (graded) information [337].
- Similarity matching, including generalization.
- Mutual interactions among fragments of representation such as rules [297], in which conclusions and conditions of multiple rules combine to produce strengthened, weakened, or new results, made possible by the lack of consistency and completeness resulting from a fragmented rule base.
- Generalization.
- Information sharing such as through top-down inheritance.
- Information sharing through percolation such as through bottom-up inheritance.
- Changing contexts and learning new rules.

The brittleness problem is pervasive. With the exception of some extremely specialized narrow domains, it shows up in reasoning in various commonsense domains. To avoid this problem, every possible scenario of combinations of conditions and actions has to be analysed beforehand and structured into a system. This is not always possible, especially for large systems. This problem makes the rule-based reasoning unwieldy in accounting for commonsense reasoning in a cognitively realistic way.

A more theoretical version of this problem has been identified [92], which points out that intuitive expertise, acquired through concrete experience, is an endangered species:

> We must resist the temptation to exalt calculative reason as personified by the computer. Instead, we must recognize that facts, rules and logic alone cannot produce either common sense, the ability to go to the heart of a problem, or intuition, our capacity to do what works without necessarily knowing why. ([92], p 27)

In a similar vein, through the study of human everyday planning activities, the importance of intuitive and subconscious actions has been recognized (or reactive and improvisatory routines), and it has been demonstrated that simple reactive routines can be used to accomplish tasks that are very difficult for traditional symbolic AI systems in terms of representational complexity. It is also evident from psychological and biological research that cognitive systems (including the human mind) must be composed of complex structures: some parts of the brain (or the nervous system as a whole) have to be able to handle rules and analytical knowledge, and other parts to handle intuitions, reactive routines, and associations. In order to combat brittleness, there is clearly a need to structure these components so that they fit together forming a functioning system, which may then have the potential of overcoming the brittleness problem.

An attempt at this, even in its most rudimentary form, will give us invaluable insight into the structure of mind and brain. This work is meant to be a step towards such an attempt.

5.2.2 Formal logics

Let us examine a few variants of rule-based reasoning. Formal logics are simple, formally defined languages, capable of expressing rules in a rigorous way [82]. The essential purpose of logics is to formalize in symbolic terms reasoning processes (such as those in mathematics and in everyday life). There are basically two types of formal logic: *propositional logic* deals with declarative sentences (propositions) that can be either true or false; *predicate logic* (first-order logic) consists of predicates which contains arguments, terms for arguments, and quantifiers constraining arguments. There is a tradition of psychological study of everyday commonsense reasoning using logic approaches. For example, Piaget attempted to build models of reasoning in which logic serves as a foundation, and in his highly influential theory, the cognitive development is described in terms of acquiring a system of mental logic gradually, with changes interpreted as the emergence of new logical competence.

However, rule-based reasoning has only had limited successes toward the grandiose goal of accounting for all kinds of reasoning. Formal logics (propositional and first-order logics) are all too restrictive to account for commonsense reasoning: they have to have all conditions precisely specified in order to perform one step of inference (and thus are unable to deal with partial, incomplete information). They take into account neither the gradedness of concepts, propositions, and rules, nor reasoning with approximate information. There is also no built-in way for similarity matching with formal logics.

For example, in FEL (see below) a rule can be as follows:

$$A\,B\,C \longrightarrow D$$

with relative weights of conditions equal to (0.3 0.3 0.3). When only A (with a confidence 1) and B (with a confidence 1) are known, we can deduce a conclusion

D with a confidence 0.6. When A is known instead with a confidence 0.6, then the confidence of D is 4.8. But in formal logic there is no way this type of reasoning can be done. Similarly, neither can classic logics deal with non-monotonic reasoning (reasoning with partial information), i.e., drawing tentative conclusions from partial, inexact information (one of the aspects of the brittleness problem we are trying to deal with) as CONSYDERR can.

5.2.3   Extensions to formal logics

There are some recent extensions to formal logics, which try to remedy some of the aforementioned problems. These extensions include *autoepistemic logic* [200], *default logic* [255], *circumscription* [191], and *variable precision logic*. Although these logics are good *normative* models and deal with partial and incomplete information, of knowledge, belief, and reasoning, they have some fundamental shortcomings (inherent in logic approaches), from the standpoint of modelling commonsense reasoning, including the lack of capabilities for dealing with:

(i)   Approximate information.
(ii)   Inconsistency.
(iii)   Reasoning as a complex, interacting process [229].

5.2.4   Probabilistic approaches

The *probabilistic approach* is a mixed symbolic/numeric approach for modelling reasoning (including commonsense reasoning). It treats beliefs (i.e., dispositions about facts) as probabilistic events, and utilizes probabilistic laws for belief combinations. Therefore, in contrast to logics, it deals with inexact information. The basic probabilistic laws used are for predicting an event based on other known events is the Bayesian rule. There have been investigations into network implementations of Bayesian reasoning [227]. With certain restrictions, some Bayesian reasoning can be carried out by local computations in a parallel network.[1]

A question one may ask is: why should beliefs (which concern often unrepeatable mental events) be combined by the law that governs repeatable probabilistic events in the physical world? The following answer is offered by [227]:

It came about because beliefs are formed not in a vacuum but rather as a distillation of sensory experiences. For reasons of storage economy and generality we forget the actual experiences and retain their mental

---

[1] However, the results so far seem to indicate that only very simple types of Bayesian reasoning can be handled efficiently by these kinds of network.

impressions in the forms of averages, weights, or abstract qualitative relationships that help us determine future actions. ([229], p 17)

Although in general we agree with this argument, it does not by itself endorse the *precise* probabilistic laws to be used in descriptive modelling. This is because commonsense reasoning does not always conform to the assumptions and the laws of probability theory, in part because of the complexity of the formal models [65]. Due to time and space resource limitations, the need for an economic organization, and the need for generalization capabilities, there are many simplifications, heuristics, and rules of thumb involved in commonsense reasoning. These approaches seem too complex to be used in human commonsense reasoning, and furthermore, probabilistic models are often computationally too complex to be implemented in 'brainlike' network models directly. Also, it is not always possible to obtain precise probability measures.

Although there are some attempts at directly applying probabilistic approaches in analysing human reasoning, which is termed *rational analysis* [12], it is proposed that human cognition should be analysed at a more abstract level, apart from implementational considerations. Anderson analysed a vast array of areas, proposing probabilistic formulations for these areas, and has had some success in accounting for experimental data. However, it seems clear that *not all* data can be dealt with in this way.

5.2.5 Fuzzy logic

Fuzzy logic is another mixed symbolic/numeric approach for rule-based reasoning and its basis, the *fuzzy set* concept, has been around for more than thirty years, and has many important theoretical results and some significant applications. The basic idea is that a particular object can belong to a set (i.e., a concept) to a certain degree, represented by a numeric *grade of membership* [337]: for each concept, there is a set of objects satisfying that concept to a certain degree, which form a subset, namely a *fuzzy subset*. This fuzzy subset contains as its elements pairs consisting of an object and its grade of membership, which represents the degree (i.e., the confidence) with which it satisfies the concept associated with the subset. Given the above, it is easy to construct a logic with which one can reason with confidence values for various concepts.

5.2.6 Fuzzy evidential logic

Let us examine a particular version of fuzzy logic, *fuzzy evidential logic* or FEL [296], which is based on connectionist models. Basically, FEL consists of a set of rules and a set of facts.

Definition 1: A *fact* in FEL is an atom or its negation, represented by a letter (with or without a negation symbol) and having a value between l and u (e.g., $[0, 1]$ or $[-1, 1]$), representing its confidence. The value of an atom is related to the value of its negation by a specific method, so that knowing the value of an atom results in immediately knowing the value of its negation, and vice versa.[2]

For example, a, $\neg a$, and x are all facts. Now we can define rules and their related weighting schemes:

Definition 2: A *rule* in FEL is in turn a structure composed of two parts: a left-hand side (LHS), which consists of one or more facts, and a right-hand side (RHS), which consists of one fact. When facts in the LHS are assigned values, the fact in the RHS can be assigned a value according to a *weighting scheme*.[3]

For example, $a \; b \; c \longrightarrow d$ is a rule, and a weighting scheme associated with it will tell us the value of d once we know the values of a, b, and c.

The above description is common to almost all fuzzy logics or production systems, but numeric calculations associated with this logic are different from either probabilistic reasoning or the fuzzy logic of Zadeh, in that a simpler and more intuitive scheme is employed [298]. We use a *weighting scheme*, which is a way of performing two things:

(i) Of assigning a weight to each fact in the LHS of a rule, reflecting its relative importance in determining the conclusion, with the total weight (the sum of the absolute values of weights) less than or equal to 1.

(ii) Of determining the value (the confidence) of the fact in the RHS of a rule (i.e., the consequent) by a thresholded (if thresholds are used) weighted sum of the values of the facts in the LHS (that is, the inner product of the weight vector and the vector of the values of the LHS facts).

When a threshold is used and the range of values is continuous, then the weighted sum is passed on if its absolute value is greater than the threshold, or the result is 0 if otherwise. When a threshold is used and the range of values is binary (or bipolar), then the result will be one or the other depending on whether the weighted sum (or the absolute value of it) is greater than the threshold or not (usually the result will be 1 if the weighted sum is greater than the threshold, 0 or -1 if otherwise).[4]

For example, a weight set for the above rule is $w_1 = 0.3$, $w_2 = 0.3$, and $w_3 = 0.4$, where each weight reflects the relative importance of the corresponding fact in determining the conclusion, and the value of d is calculated by the weighted sum (or inner product) of the values of the facts in the LHS (i.e., the confidence values of the conditions): $d = w_1 a + w_2 b + w_3 c$.

---

[2] We will adopt a generic confidence measure as the value of a fact. We will later further characterize this measure.

[3] When the value of a fact in the LHS is unknown, assign a particular value representing *unknown*, say zero, as its value.

[4] This weighting scheme can be generalized, as will be discussed later on.

We will define the notion of a theory:

Definition 3: A theory is a 4-tuple: ⟨A, R, W, T⟩, where A is a set of facts, R is a set of rules, W is a weighting scheme for R, and T is a set of thresholds each of which is associated with one element in A. I is a set of elements of the form (f, v) (where f is a fact, and v is a value associated with f), representing initial input.

One issue that needs to be clarified is how to handle multiple rules reaching the same conclusion (i.e., the same fact is at the RHS of more than one rule, and they are all activated), in which case we have to combine the results somehow. One simple way is to take the MAX of all values from various rules regarding the particular fact concluded. As a matter of fact, this is the *best* way to do it, given all the constraints and considerations we have (when a system is consistent), as will be elaborated later when we discuss rule interactions.

Definition 4: A *conclusion* in FEL is a value associated with a fact, calculated from rules and facts by doing the following:

(i) For each rule having that fact in its RHS, obtain conclusions of all the facts in its LHS (if any fact is unobtainable, assume it to be zero, or any value used for representing *unknown*), and then calculate the value of the conclusion in question using the weighting scheme.

(ii) Take the MAX of all these values associated with that fact calculated from different rules or given in initial input.

Definition 5: A rule set is said to be *hierarchical*, if the graph depicting the rule set is acyclic; the graph is constructed by drawing a unidirectional link from each fact in the LHS of a rule to the fact in the RHS of a rule. Making a rule set hierarchical avoids circular reasoning.

Now FEL can be defined as follows:

Definition 6: A *fuzzy evidential logic* (FEL) is a 5-tuple: ⟨A, R, W, T, I⟩, where A is a set of facts, the values of which are initially assumed to be zero (or any value used for representing *unknown*), R is a set of rules, W is a weighting scheme for R, T is a set of thresholds each of which is for one rule, and I is the initial condition: a set of elements of the form (f, v), where f is a fact, and v is a value associated with f and C is a procedure for deriving conclusions (i.e., computing values of facts in the RHS of a rule in R, based on the initial condition I and previously derived conclusions) compatible with the rule set R and the initial condition I.

We want to differentiate FEL into two versions: FEL_1 and FEL_2, which differ in their respective ranges for values associated with facts.

Definition 7: FEL_1 is FEL when the range of values is restricted to between 0 and 1 (i.e., $l = 0$ and $u = 1$), and the way the value of a fact is related to the value of its negation is:

$$a = 1 - \neg a$$

for any fact a.

Definition 8: FEL_2 is FEL when the range of values is restricted to between -1 and 1 (i.e., $l = -1$ and $u = 1$), and the way the value of a fact is related to the value of its negation is:

$$a = -\neg a$$

for any fact a.

The difference between FEL and conventional fuzzy logic is that FEL uses the weighted-sum computation (borrowed from connectionist models) for combining evidence *cumulatively* to reach a conclusion from a rule.[5] Yet another difference is in terms of ways of evidential combination. With MAX/MIN Zadeh's fuzzy logic does not accumulate evidence (only using the evidence with the highest grade of membership), which is nevertheless important for commonsense reasoning [297] on the basis of the human data [73]. FEL provides a scheme for plausible reasoning, which accumulates evidence by using a weighted-sum computation, and it has a sound basis in logic in terms of reducing to classical logics when all values are zeros or ones, because it reduces to Horn clause logic. It is natural for implementation in connectionist models, for it adopts weighted-sum computation. An alternative way of looking at it is that it provides a semantics for weighted-sum connectionist models, by viewing weighted-sum connectionist models as combining evidence from the condition part of a rule and calculating the certainty of the conclusion. Unlike conventional fuzzy logic, in FEL we do not have to know all grades of membership in order to perform one inference step: if we have no information regarding the value of a fact, we simply assume it to be zero. This way we can easily allow partial match to a rule, and draw tentative conclusions in a nonmonotonic fashion. For example, we can have the following rules in FEL:

$$A \longrightarrow C \text{ weight} = (0.8)$$

$$A B \longrightarrow D \text{ weight} = (0.5, 0.5)$$

assuming the weight distributions already take into account the certainty of rules. In this case, when only A is known (with confidence 1), C is

---

[5] Alternatively, we can view FEL as providing a semantics for the weighted-sum connectionist models. Thus FEL provides a foundation for carrying out fuzzy rule-based reasoning in connectionist networks.

concluded.   When A and B both are known (with confidence 1), D is concluded, because D is activated more strongly than C. For example, suppose A = bird, B = penguin, C = fly, and D = not fly. Given penguins (and birds) in the input, we can reliably conclude that they do not fly; on the other hand, given birds alone, we can conclude that they do. This flexible handling results from the parallelism and the numerical weighting scheme of the connectionist model. In typical logic-based or otherwise rule-based systems, one rule and only one rule will be picked to fire at a time, but in CONSYDERR all reachable conclusions can be reached, and the final outcome is the combination of these partial results. Moreover, it is also interesting that FEL satisfies all relevant axioms of various recent nonmonotonic systems, such as *reflexivity, left transitivity, right weakening, cut, reciprocity*, and *distribution* [297]. By using weighted-sum computation and allowing the total weight of a rule to be less than one, FEL rules can be uncertain (graded), making it more expressive of complex real world situations. With such rules, the confidence for conclusions reached along a chain of inferences weakens along the way, corresponding better to human commonsense reasoning. FEL is aimed at reducing computational complexity in probabilistic approaches (as mentioned above) by simplifying the probabilistic formulas. In terms of probabilistic reasoning, FEL weights can be viewed as conditional probabilities, and values of conditions and conclusions can be viewed as the probabilities of the corresponding events. Thus, FEL can be viewed as probabilistic reasoning under the assumption of independent evidence (i.e., pieces of evidence are exclusive and exhaustive).[6]   In sum, FEL deals with multiple aspects of the brittleness problem. Although FEL can be presented in a symbolic framework, it differs from symbolic approaches (including formal logics) in that it does not perform inferences sequentially and symbolically, and it does not require indexing, retrieving, reorganization, or updating of large databases[7] as needed with symbolic logics, and can utilize mutual interactions. The reasoning process is thus more cognitively realistic [299].   The main characteristics of FEL are that:

- It uses addition and multiplication (which are natural for connectionist networks), instead of MAX and MIN.
- It allows partial matching and accumulation of evidence.
- It distributes weights to different pieces of evidence, and thus can emphasize certain things and disregard others as appropriate.
- It treats, in a uniform framework, not only linguistic vagueness but also other kinds of inexactness.
- It accounts for the phenomenon that, in a chain of reasoning, the certainty of conclusions weakens along the way.

[6] However, because of this simplifying assumption, we may lose some finer details of the probability theory, including interactions among different conditions in a rule.
[7] The computation is done in place, without the hassle of moving data around, by directly connecting related facts (related by rules) together so that only local computation is necessary. This results in massive parallelism (as will be discussed later).

5.2.7   Representation and reasoning with rules

This brings up another issue, knowledge representation, which has been a central issue in AI. Many different forms of representation have been explored, including rule- (and logic-) based representation. More representational forms have been explored in AI for modelling knowledge and inferential structures now than ever before in philosophy, psychology and other relevant fields. The sudden influx of new ideas is certainly attributable to the use of computers both as a highly efficient tool for implementing and testing various forms of representations and as a uniquely powerful metaphor for cognitive agents when constructing representational frameworks in modelling such agents. However, early-day knowledge representation research emphasized static (symbolic and declarative) representations in a static way, concerned mainly with the form of representations, with little work exploring how the static pieces of knowledge can be utilized to draw useful conclusions dynamically in a principled way. Representational frameworks were demonstrated by a few often captivating but superficial examples. Little theoretical analysis and fundamental work was done. When the pitfalls of this direction became obvious, more emphasis was shifted to inferences that can be performed based on a given representation, instead of how pieces of knowledge are actually organized and stored. This is certainly a healthy trend, given that various representations are more or less alike, and many seem to be only notational variants of each other. Some theoretical results can be established that determine the relationship between different representational frameworks and/or equate them in some ways. On the other hand, *inferential structures*, which determine how inferences are made, are much more volatile and are more controversial in terms of their characters. Although some progress has been made that leads to better understanding of *inferential structures* (that is, what kinds of inference are allowed or prohibited) by means of traditional and non-traditional logics as discussed above, there is little principled understanding of what is underlying such inferential structures. In other words, most research work being done is concerned with *surface features* of inferential structures rather than the underlying deep causes or principles. To see this, it suffices to mention the fact that inferential structures are mostly characterized as a set of axioms, instead of by deep causes; the inferential steps are described in purely syntactic terms instead of involving semantic and/or implementational considerations. Moreover, inferential structures are completely separated from the knowledge represented; that is, inference rules (and the mechanism that uses these inference rules to draw conclusions) form a separate component, independent of the knowledge base. In fact, the question of what inferences to make depends on a large number of factors. Besides the purpose of inferencing and requirements of speed, accuracy and generality (which depend on particular contexts), the underlying physical machinery is an important consideration, because the substrate constrains the phenomenon. Here, since we are more interested in the generic reasoning capability, the generic characterization of the interaction of substrate (implementations) and inference structures are the main issue for us. Studying substrates helps to answer questions such as how certain inferences are facilitated and how certain inferences are prohibited. In other words, one might obtain a (substrate-based) structural explanation for reasoning functions. At the same time, one might obtain a (function-oriented) teleological explanation for substrate organizations. The mutual impact of functions and organizations and organizations upon functions can be explored to the benefit of greatly increased understanding of inferential structures in human commonsense reasoning. By utilizing the parallel between substrate structures and reasoning functions, the task of defining inferential structures for approximate commonsense reasoning, which goes far beyond classical logics, can be less underconstrained and better motivated, and various design options can be explored. Competence and performance are unified in one structure. Mappings from functions to structures and back can be made out in the interest of building cognitively plausible, commonsensical intelligent systems. In sum, the problem with symbolic rule-based reasoning, or symbolic models in general, is that they tend to be far removed from biological implementations (although they may not be so necessarily) and consequently they may not be able to capture the intrinsic properties stemming from substrate constraints that are indispensable to the understanding of commonsense reasoning processes. In direct contrast to symbolic approaches, we believe that inferential structures are inseparable from substrate organizations in the study of knowledge representation. In considering substrate organizations, one important constraint is *local computation*. The method of global data storage, retrieval, and reorganization requires such an overhead that it has been considered implausible as a model for cognitive architectures [297]. A better method is local computation; that is, computation (i.e., inferences) is done right where data (i.e., stored knowledge) are. Pieces of knowledge are to be distributed in a vast network linked up by inferential connections between them; i.e., inferential structures can be one and the same as (isomorphic to and embodied in) substrate organizations. Computation necessary for such a network is carried out locally, within nodes and along links, without global control, monitoring, or storage. Putting the above two points (i.e., the substrate structure and the local computation) together, instead of separating knowledge to be represented and inferences that can be drawn from it into two distinct parts, as is customarily done in symbolic AI, we can integrate the two and view them as the two aspects of the same process. Due to improved efficiency, such a method may better tackle the brittleness problem (below we try to realize the above desiderata).

5.3   Case-based reasoning

The idea of case-based reasoning was proposed as a strong antithesis to the more traditional rule-based reasoning. *Cases* refer to concrete scenarios and episodes that can be used in problem solving. Case-based reasoning deals with an important aspect of the brittleness problem: similarity matching when there is no exact rule. Case-based reasoning is characterized [257] as performing the following steps:

(i)   INPUT, i.e., receive data.
(ii) INDEX, i.e., find relevant pointers.
(iii) RETRIEVE, i.e., find similar cases based on indices.
(iv) ADAPT, i.e., modify the retrieved cases for the current situation.
(v)   TEST, REPAIR, and STORE, i.e., try out the solution and store it into the case base after it is tuned into a correct form.

This entire process is construed to be symbolic, sequential, and iterative. They present many interesting examples of systems performing case-based reasoning, in planning and problem solving. One example used to support the idea of case-based reasoning is as follows: when an architect starts a new design for a client, he/she does not go back to first principles and try out all possible combinations. Instead he/she recalls past similar plans and modifies them to fit his/her current needs [128].

There are the following reservations about the purely symbolic case-based approach. First of all, rules (i.e., abstract knowledge) are one of the most important cognitive mechanisms, although no doubt there are other kinds of knowledge (such as cases). Generally speaking, only when there is no rule directly applicable, should cases (or analogous knowledge in general) be used. For example, when there is a rule available such as 'if a place is warm, flat, and with enough fresh water supply, **then** it can be a rice-growing area', we only need to apply the rule in deciding if an area can be a rice-growing area or not. When there is no such rule, or the conclusion of the rule is indecisive, we need to apply analogous knowledge (or cases). Of course, such rules may not be context-free and work in isolation as often seen in today's expert systems. Even in existing case-based reasoning systems, various forms of rules are used at various points, which is more evidence for the primary role of rules in cognition. There ought to be, therefore, both rules and cases (similarity matching of analogous knowledge) in the system, and rules should take precedence. Secondly, similarity matching used in retrieving cases, at least in most situations, should be done at a lower level; that is, matching should be an intuitive, holistic process as there are multiple levels/components in cognitive systems [93].

It seems that in human reasoning in the above described situations, non-explicit, subconceptual, and intuitive modes prevail. Only when unexpected events occur and usually successful routines are interrupted, does explicit conceptual reasoning take over to reason from rules and first principles to find a solution for unusual situations [93]. An even clearer example is as follows: in daily commuting to work, one does not have to rethink the route; rather it is a routine to get into the car, drive onto the streets, and follow the usual route to work. However, when the route is unexpectedly closed, then one has to think and reason explicitly about the alternatives. The question is: how can case-based reasoning better model non-explicit, subconceptual reasoning, in addition to explicit reasoning?

Thirdly, similarity matching should be done in a massively parallel fashion and spontaneously (at least for some portions of it), without incurring huge computational overhead, because, for commonsense reasoning, the matching and adaptation process cannot be slow and deliberative [297]. Simpler algorithms are thus preferred. Rules, in terms of their form, can be a useful representational framework for both concrete and abstract knowledge. CONSYDERR adopts a simple, uniform representation that encompasses both cases and rules. In sum, it seems that although case-based inference (or the use of analogous knowledge in general) is very important for commonsense reasoning, it is supplementary to abstract knowledge encoded in rules. In other words, rules are primary, and concrete analogous knowledge is secondary, and there should be both rules and cases (with rules taking precedence).

Finally, rules and similarity matching should preferably be integrated into a unified framework. The counterpart of case-based reasoning in psychology (which often serves as the motivation for case-based reasoning) is the study of similarity and analogy. A detailed model has been provided of how similarity figures in decision making processes, i.e., the case matching process. Here the contrast model [319] for the computation of the similarity between two given concepts was adopted. Based on this similarity measure, and on the basis of that, the authors explain a number of phenomena in human decision making such as the *conjunction fallacy*, the *base rate effect*, and *causal effects of base rate*. Their model is based on computing the overall similarity between the properties of the concept representation and the properties of the observed objects (utilizing the frame representation). This work thus represents a step toward a detailed cognitive model of similarity and its role in approximate commonsense reasoning (as opposed to formal reasoning).[8]

There is a computational model for analogical reasoning in which the mechanism for accomplishing this process is unified with the basic rule-based representation and the search method. Thus the system deals with both the use of analogical knowledge and the direct application of rules: when directly applying rules fails to solve a problem, a number of factors (including identical concepts and chaining of subgoals) lead to the activation of an analogous problem which has a stored solution. After establishing correspondence between the two problems, performing analogous actions in the target problem is enabled, and a new solution is thus formed. The question is whether this particular rule-based architecture can be scaled up to problems of a realistic complexity; a massively parallel connectionist framework might be better off in avoiding being bogged down by combinatorial explosion. This system mixes rule-based reasoning with

---

[8] It also somehow constitutes a step that can lead to feature-based connectionist representation, because the properties used can be readily implemented as features in connectionist networks.

similarity-based (case-based) reasoning, but notably both are carried out in very explicit fashions. A problem with such explicit representation and reasoning in general is, as mentioned earlier, that they do not capture the subconceptual, holistic process that seems most likely to be underlying similar human reasoning. As a result, the system is still brittle and unwieldy in terms of representational and computational complexity.

There have been repeated arguments for a holistic representation of similarity. From a different perspective, it has been argued that symbolic representation cannot account for intuition and situation dependent reasoning [93], and that this leads AI to study only deliberative rationality (analytical reasoning). Rationalized thinking tends to be situation independent, abstract symbol manipulation, which could be well handled by symbolic paradigms, as implied by the *physical symbol hypothesis* [212]. But on the other hand intuitions are situation dependent (or context sensitive), well grounded in experience, and holistic in nature. Deliberative rationality is based on symbol manipulation and therefore is suitable to be modelled by symbolic systems. However, reason or rationality has to be rooted in intuition, otherwise it will not qualify as intelligence. It has been convincingly argued [93] that the holistic, holographic nature of human intuition distinguishes it from rationalized thinking. It was suggested that holistic, holographic *similarity* plays a large role in intuition. Due to their distributed nature, connectionist models might be able to better model such similarity that so far the symbolic paradigm has failed to capture. In the light of all of the above arguments, it is plausible that while rule applications can be modelled at a higher level by a symbolic process, which represents explicit conceptual knowledge, fine-grained similarity matching should be modelled at a lower level by an informal, holistic, structureless process, which embodies intuition.

5.4 Connectionism

Connectionist networks are fine-grained, massively parallel computational systems that account naturally for the parallelism in human reasoning, including commonsense reasoning. Throughout its history, connectionism has always had as one of its main goals the modelling of reasoning and cognition, and intelligence. Connectionism aims at developing systems that can learn from experience and that are capable of generalization from data (and can thus avoid brittleness), by using numeric functions and by learning incrementally. Thus, connectionism can potentially be a good antidote to brittleness in symbolic commonsense reasoning models. Despite its promise, it requires an effort to apply the connectionist paradigm *directly* to the study of commonsense reasoning. In this section we will look into some efforts at understanding reasoning, rules, and high-level cognition within the connectionist framework. We will see how it fares as a remedy to the brittleness problem.

The view has been put forward [290] that there are two distinct types of process underlying cognition (which correspond to the conceptual and subconceptual levels respectively), both need to be taken care of, and connectionist models are good for modelling one of them. The *conceptual level* possesses three characteristics:

(i)   Public access.
(ii) Reliability.
(iii) Formality.

It can thus be modelled by symbolic processing. Smolensky reaffirms (as numerous philosophers from Plato to Kant have done) the appropriateness of modelling such conceptual knowledge by symbolic processes, as customarily done in symbolic AI, but on the other hand states that skill, intuition, individual knowledge, and so on, are not expressible in linguistic forms and do not conform to the three criteria prescribed. It seems futile to try to model this kind of knowledge in symbolic forms for the following reasons:

(i)   The resulting systems from the symbolic approach are often too brittle and inflexible, even from the symbol processing viewpoint.
(ii) There are currently major difficulties unsolvable in some important domains, such as natural language understanding, planning, and learning; symbolic systems are far from comparable to human performance, despite tremendous efforts put into them. Thus it seems that symbolic processing is not the appropriate framework.
(iii) Symbolic systems are too far removed from biological implementations and thus they may not be able to capture some of the intrinsic properties of natural intelligence generated by substrate constraints.

Hence, according to the above view, some cognitive capacities, such as skill, intuition, and individual knowledge, constitute different kinds of capacity (that is, subsymbolic processing at the *subconceptual level*) and may be better dealt with by the connectionist paradigm using distributed connectionist models [290], which can then be used to address the brittleness problem.

5.4.1 High-level connectionist models

Recently there have been a number of high-level connectionist models proposed. For example a connectionist knowledge representation system [87], μKLONE, is presented, in which reasoning is carried out by constraint satisfaction through minimizing an error function (which is proportional to the outstanding constraints). By constructing the function in a way that captures the underlying relations between logical formulas, this process can produce an optimal (or near-optimal) interpretation of the inputs. The actual process of energy minimization is done by the *Boltzmann machine* settling procedure. One problem with this approach is that the settling process is extremely slow, and may not produce the best results. A serious limitation, however, is that not all logical relationships between sentences and all inference modes can be captured in error functions. As a matter of fact, most of the commonsense reasoning cannot be done by this error minimization procedure in any obvious way [299].

A hybrid system of marker passing and connectionist networks has been presented [139] to augment symbolic planning systems in their ability to pick out the right move by priming the relevant concepts. The role of the connectionist network is that of association, and therefore there are no rules and no variable binding capabilities. Although this works well in a particular situation, for more complicated tasks we have to go beyond mere associations [296] and into more elaborate mechanisms for more systematic reasoning within the framework of connectionism.

5.4.2   Connectionism and rules

Despite all of its faults, explicit rule-based reasoning is still needed in connectionist models. This need can be argued for in the situation in which one can learn and manipulate rules consciously with *ease* and thus there ought to be an explicit rule representation that can be explicitly manipulated, as otherwise there is no way to capture that degree of manipulability. Another situation is that in which conventional distributed connectionist models are identified with sub/unconscious processing, due to the nature of implicit representation in conventional distributed connectionist models, and evidence is given for explicit, conscious rule applications in cognition, contrary to the nature of sub/unconsciousness. Therefore the conclusion is drawn that there should be a separate level, or even a separate component, that can account for explicit conscious rule application in addition to conventional distributed connectionist processes. Eight criteria have been presented for the existence of rules in cognition [288], and a detailed analysis of experimental results shows that the eight criteria can be met by various data. Thus the conclusion is drawn that rules are a necessary part of certain reasoning.

There are many connectionist networks which implement rules directly in connectionist networks. But in most of these networks parallelism is lost in some way, either because of the matching process of hardwired rules or a centralized working memory. Systems developed more recently [6] can implement the complete first-order predicate calculus in connectionist models using complex network structures. Such work is concerned with the computational issues, not the cognitive issues (including modelling commonsense reasoning). Although some of them handle partial matching, most of these systems do not deal with uncertain, graded information [158].

5.4.3   Connectionism and cases

Connectionist models are inherently similarity based: responses to inputs are determined by the similarity of the input to the prototypical cases in the previous training data, which can be characterized as performing feature clustering based on similarity or as performing statistical inferences. The distributed representation in networks was analysed [141] to show how similarity-based generalization can occur naturally, which can help to deal with the brittleness problem. While these above kinds of similarity tend to be unstructured and fine grained, more structured types of similarity have been explored by others [36]. In this study an attempt at utilizing case-based reasoning in connectionist networks was described, for the purpose of combating the brittleness that is undermining both purely symbolic rule-based reasoning systems and connectionist rule-based systems. Thus the system described, COMPOSIT, is particularly relevant. The system adopted from their original rule-based connectionist system discussed before, has multiple configuration matrices (CMs), each of which contains a case for short-term processing. A relatively small set of gateway CMs provides the interface between short-term processing and long-term memory, which encodes cases in connection weights. Cases in non-gateway CMs compete to have their contents copied into gateway CMs, where they can cause cases similar to it to be retrieved from long-term memory, and some symbol substitutions take place after the retrieval to adapt the cases to the current situation. One drawback is that gateways and copying/substitution mechanisms used unnecessarily hamper parallelism. This system does not handle partial match and uncertain (approximate) information.

5.4.4 Four criteria in integrating rules and similarities

The above discussion points to the need for integrating rule-based and similarity-based reasoning in a better way. To help to choose the most plausible cognitive architecture that integrates the two, rule-based and similarity-based reasoning, four criteria can be hypothesized based on the preceding discussions and arguments:

- *Direct accessibility of concepts.* To model conceptual-level reasoning and especially explicit rule-based reasoning, concepts (and even reasoning processes) should be directly accessible and linguistically expressible; that is, they can be accessed directly, without intermediate steps and additional matching/extraction components. This requirement calls for explicit representation for concepts and basically rules out distributed representation, for a concept represented by a distributed pattern is not *directly* accessible. This requirement leaves us with two options: purely symbolic representation and localist connectionist representation.
- *Direct inaccessibility of similarity matching.* As argued before, at least certain types of similarity matching are done subconceptually, in a holistic way. The actual matching processes may not be accessible directly and conceptually. This is important, because holistic operation entertains a host of properties that other modes of operation lack, such as computational

tractability, context sensitivity, and massive parallelism [297]. A distributed representation seems to be the only viable way to meet this aspect.

- *Linkages from concepts to features.* Once a concept is activated, its essential features are also activated, either explicitly (conceptually) or primed subconceptually somehow. This (explicit or implicit) activation of features is important in subsequent similarity matching and other uses of the information associated with the concept.

- *Linkages from features to concepts.* Once all or most of the features of a concept are activated (implicitly or explicitly), the concept itself should be activated to 'cover' these features. This corresponds roughly to the categorization process [289].

These four criteria strongly favour a two-level architecture, with one level for explicit conceptual representation and reasoning and the other level for features and implicit, subconceptual processes. Such a framework containing both types of reasoning enables the exploration of their synergy, and thus can serve as a basis for building more flexible and powerful systems incorporating different reasoning modes. One question in such a framework is how to model the subconceptual, holistic similarity matching process within such a two-level framework. Distributed representation seems to be the best choice we have [143]. Other researchers [341] have demonstrated to various degrees the (holistic) symbolic processing capabilities of distributed representation.

The above conjectures lead naturally to the CONSYDERR architecture [299], a connectionist architecture for knowledge representation and reasoning. Briefly, CONSYDERR consists of two levels: CL and CD. CL is a connectionist network with localist representation, or roughly reasoning at the conceptual level [290]. Rules are represented in CL as links between two nodes representing the condition and the conclusion, respectively, in accordance with FEL (i.e., directly implementing FEL). FEL can handle a superset of Horn clause logic and Shoham's logic (or *causal theories*), so that it can fully accommodate traditional rule-based reasoning and capture commonsense causal knowledge. Moreover, it is capable of approximate and cumulative evidential reasoning and works with partial and uncertain information. Unlike Horn clause logic, it can deal with negative as well as positive evidence. Note that it can also handle *variable binding* by utilizing a particular formalism [295]. Thus, inferential structures for approximate reasoning are enmeshed in the localist network. On the other hand, CD is a connectionist network with distributed representation, roughly corresponding to reasoning at the subconceptual level. Concepts and rules are diffusely represented by sets of (feature) units overlapping each other. The amount of overlapping of two sets of these units representing two different concepts is proportional to the degree of similarity between these two concepts. This is a *similarity-based* representation, in which units can be subconceptual features, perceptual primitives, internal goals, or affect states. Concepts are 'defined' in terms of their similarity to other concepts in these

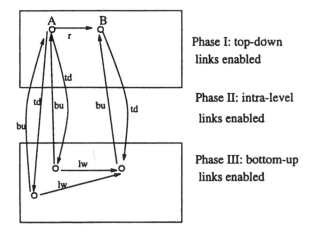

Phase I: top-down
links enabled

Phase II: intra-level
links enabled

Phase III: bottom-up
links enabled

Figure 5.1. A two-level architecture: CL is the top level and CD is the bottom level.

primitive representations. We will utilize these primitives only as a substratum for similarity-based representation of higher-level concepts. We link the localist network (CL) with this distributed network (CD), by linking each node in CL representing one concept to all the nodes in CD representing the same concept, and assign them appropriate weights (see figure 5.1).

Those cross-level links (*td* and *bu*) are moderated by a latch mechanism. The rule links in CL (labelled as *r*) are duplicated (diffusely) in CD (labelled as *lw*). The interactions of the two components are in fixed cycles: First the latch opens to allow the activation of CL nodes to flow into corresponding CD nodes (because of the similarity-based representation, the CD representations of all the concepts that are similar to the activated ones are partially activated). Then the two parts start settling down on their own simultaneously (through applying rules represented by links). Finally the latch opens to allow the activation of nodes in CD to flow back into CL to be combined with the activation of corresponding CL nodes (to accomplish both similarity matching and rule application). Also, through the top-down and bottom-up flow, the two types of inheritance (top-down inheritance and percolation as mentioned above) can be handled implicitly. Thus, CONSYDERR addresses all of the major aspects of the brittleness problem as listed earlier.

After starting to receive input data, the CONSYDERR system operates in three fixed cycles, the top-down phase, the settling phase, and the bottom-up phase. These phases can be repeated to continuously track inputs.

Definition 9: In the top-down phase (figure 5.1), the computation is as follows:

$$x_i = \max_a ACT_a \qquad (5.1)$$

where a is any node in CL that has $x_i \in CD_a$.

In the settling phase, the computation is as follows:

$$ACT_a = \sum W_i * I_i \qquad (5.2)$$

$$x_i = \sum w_i * i_i \qquad (5.3)$$

where W_is and w_is are rule strengths (weights) measures, and I_is and i_is are the activations of related concepts or features (premises or logical predecessors).

In the bottom-up phase, the computation is as follows:

$$ACT_b = \max\left(ACT_b, \sum_{x_i \in CD_b} \frac{x_i}{|CD_b|} \right) \qquad (5.4)$$

where b is any node in CL.

An (oversimplified) example drawn from a human reasoning protocol involving questions and replies [73] is as follows:

Q: Is Chaco cattle country?
R: It is like western Texas, so in some sense I guess it's cattle country.

Here because there is no known knowledge (or no applicable rules), an uncertain conclusion is drawn based on similarity with known knowledge (rules). We can restructure the protocol in accordance with rules and similarities, to straighten out the reasoning:

Definition 10: Western Texas is cattle country.
Chaco is similar to western Texas (in some relevant aspects).
So Chaco is cattle country.

Applying the three phases to the 'Chaco' example: first the node representing 'Chaco' is activated; the top-down phase will activate the CD representation of 'Chaco' and activate partially the CD representation of 'western Texas' based on their similarity. Then in the settling phase, rules (links) take effect and this amounts to applying in CD the rule: *western Texas is cattle-country*, so the CD representation of 'cattle country' is partially activated. Finally in the bottom-up phase, the partially activated CD representation of 'cattle country' will percolate up to activate the 'cattle country' node in CL. The result can be read off from CL (figure 5.2).

Let us compare, in the light of the four criteria, some existing high-level connectionist models that try to integrate rules and similarities. Some of these models [36] do not satisfy the second criterion, and some do not satisfy the first criterion [315]. On the other hand, CONSYDERR is exactly the framework needed, with a two-level dual representation that contains both localist and distributed representations. The localist network encodes rules and performs rule-based reasoning *explicitly* with directly accessible

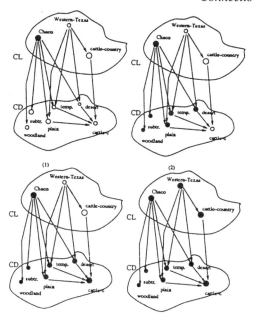

Figure 5.2. The reasoning process for the Chaco protocol: (1) receiving inputs, (2) top-down, (3) settling (rule application), (4) bottom-up.

conceptual representation, which is essential to commonsense reasoning tasks, and the distributed network encodes similarities *implicitly* with feature-based representation. Similarity matching is accomplished through a top-down/bottom-up cycle that directs the information flow in a circular way; thus the process is not directly accessible conceptually. Compared with these other existing high-level connectionist systems, CONSYDERR has the following advantages. The overall architecture of dual representation is a principled way, based on the dichotomy of conceptual and subconceptual processing, of exploring the synergy between rule-based reasoning and connectionism, or between rule-based components and similarity-based components. This integration on the one hand adds continuity (embodied in similarity-based connectionist networks) to the discretized rule-based system, so that it can better model continuous thought processes. On the other hand, it adds structures (namely rules, as well as variables and bindings) to a structureless, connectionist network, giving it the rigour, precision, and directedness which it needs to model some cognitive tasks. Based on the combination of similarity-based reasoning and rule-based reasoning, many difficult patterns in commonsense reasoning emerge without being explicitly put into it [299]. The synergy helps to deal with the aspects of the brittleness problem identified earlier, such as partial information, similarity matching when there is no exact rule, inheritance, and rule interaction, with only very simple representation. These issues are not treated in depth here but are discussed elsewhere in more depth [301]. The architecture also provides a unified framework for both conceptual reasoning and subconceptual intuitive reasoning (although in a limited way), so that their synergy can be utilized and brittleness can be alleviated.

Another important characteristic of CONSYDERR is that links between nodes that represent pieces of knowledge (concepts, propositions, and features) are used to direct inferences based on rules (in FEL). Thus the inferential structure of the system is captured right in the representation, and substrate constraints are incorporated into reasoning, as demanded in the earlier discussion. Yet another important characteristics of CONSYDERR is the combination of conceptual level and subconceptual level reasoning. By combining the two types of reasoning mode, we are able to explore the synergy resulting from the interaction between them and between rule-based reasoning and similarity-based reasoning. This synergy helps to deal with aspects of the brittleness problem identified earlier, such as partial information, no exact matching, property inheritance, rule interaction, with only very simple representation as explained above.

There are some commonalities and differences between the philosophies behind CONSYDERR and case-based reasoning. The commonality is evident: both approaches utilize similarity between the current situation and previously known situations to come up with a plausible conclusion, especially if we look at one of the applications, GIRO. The differences between this approach and the approach taken by case-based reasoning are the following:

- Rules are the basic coding mechanisms for both concrete and abstract knowledge in CONSYDERR (i.e., abstract knowledge) (the word *rule* is used here in a loose sense, denoting anything that can be expressed in some sort of logic). CONSYDERR adopts a simple, uniform representation that encompasses both cases and rules. There are other kinds of knowledge. Only when there is no rule directly applicable, will cases (or analogous knowledge in general) be used. For example, a rule is: **if** a place is warm, flat, and with enough fresh water supply, **then** it can be a rice-growing area. When there is such a rule available, we only need to apply such a rule in deciding whether an area is a rice-growing area; only when there is no such rule, or the conclusion of the rule is indecisive, may we apply analogous knowledge (or cases). Of course, such rules are not context free and work in isolation as often seen in expert systems. Even in existing case-based reasoning systems, various forms of rules are used, which is more evidence for the primary role of rules in cognition.
- In CONSYDERR, similarity is not only captured in link weights but also directly in matching cases coded in CL by the top-down and bottom-up flow via the similarity-based distributed representation in CD.
- There are both rule application and similarity matching in CONSYDERR, and rules take precedence.

- Similarity matching is done in a massively parallel fashion, and thus is very efficient. It is done spontaneously, without incurring huge computational overhead, because, for commonsense reasoning, the matching and adaptation process cannot be very analytical and deliberative. Simpler algorithms are thus preferred.
- Similarity matching should be done at a lower level, that is, matching should be an intuitive, holistic process [93].
- Rules, in terms of form, can be a useful representational framework.

In short, although case-based inference (or the use of analogous knowledge in general) is very important for CONSYDERR, it is supplementary to abstract knowledge (encoded in rules). In other words, rules are primary, and concrete analogous knowledge is secondary, and they are integrated into a unified framework. We can also compare CONSYDERR with connectionist case-based reasoning [36], and notice some differences:

- CONSYDERR has a higher degree of parallelism. There are no gateways and complex copying and substitution processes that may hamper parallelism.
- Matching in CONSYDERR is more automatic (spontaneous). There is no special matching mechanism (although this potentially limits the matching capability).
- One uniform representation (i.e., rules, in terms of form) takes care of both rules (in terms of contents) and cases in the same system.

Overall, the two different systems represent two different attacks on the same problem, and the two approaches might converge somehow. On the other hand, there are also strong contrasts between CONSYDERR and pure symbolic rule-based reasoning:

- CONSYDERR handles similarity-based reasoning in a natural and inexpensive way.
- CONSYDERR utilizes the synergy between two types of representation and reasoning, so that several difficult patterns in commonsense reasoning can be generated with ease.
- CONSYDERR executes rules in a massively parallel fashion, and thus deals with mutual interactions among rules.
- CONSYDERR handles uncertain, approximate, and graded information, which most symbolic models choose to ignore (except fuzzy logic and probabilistic reasoning).
- CONSYDERR also handles situations where only partial, incomplete information is available.

Because of these above advantages, CONSYDERR is better in alleviating the brittleness problem. In comparing CONSYDERR with connectionist rule-based reasoning systems, most of these systems do not deal with similarity matching as in the CD part of CONSYDERR, and consequently they are unable to utilize the synergy between the two types of representation essential to CONSYDERR; thus they are unable to do more than what a typical symbolic rule-based system is capable of, besides massive parallelism (but they do not deal with reasoning as a complex process). Therefore, they suffer the same brittleness problem as symbolic systems.

5.5   Conclusions

This chapter brings together several paradigms of artificial intelligence, and examines some representative models in each of them.   These paradigms include rule-based reasoning, which is a predominant paradigm of symbolic AI, case-based reasoning, which is also mostly cast in symbolic forms, and connectionist models.   The characteristics of each of these paradigms are identified and/or critically analysed.   This chapter advocates the integration of rule-based reasoning and similarity-based reasoning (as embodied in both case-based reasoning and connectionist models). Various existing models have proven their respective advantages and usefulness, despite various shortcomings associated with each of them. They have complementary characteristics and can work together synergistically.

This integration is related to some philosophical and psychological arguments regarding different types of cognitive process and their respective properties.   Specifically, Smolensky posited the distinction between the conceptual level and the subconceptual level and Dreyfus emphasized the difference between deliberative/analytic thinking and intuitive thinking. Rule-based reasoning tends to capture the conceptual-level processes, while similarity-based reasoning can help to capture the subconceptual-level, intuitive processes.

Four criteria are hypothesized, regarding the integration of conceptual and subconceptual processes: direct accessibility of concepts, direct inaccessibility of similarity matching, linkages from concepts to features, and linkages from features to concepts.   These four criteria which jointly determine a certain range of possibilities in terms of architectures for integrating conceptual and subconceptual representation lead naturally to a two-level architecture, with one level for explicit conceptual representation and reasoning and the other level for implicit, subconceptual processes.   Such architecture facilitates the utilization of the synergy between the two types of process which helps to eliminate brittleness in AI systems for commonsense reasoning. Such a two-level architecture not only captures the cognitive distinction of conceptual and subconceptual processes, so that models of commonsense reasoning processes can be cognitively more plausible and realistic, but also enables the utilization of the synergy resulting from the interaction of the two levels, so that such models can eliminate the brittleness problem of typical rule-based systems caused by their one-sided, over-explicit way of reasoning, while maintaining a certain precision and rigour that are necessary [299]. CONSYDERR integrates the two processes with a dual localist and distributed representation, as well as rule-based reasoning and similarity-based reasoning. Firstly, this integration is novel and is based on the dichotomy of conceptual and subconceptual (intuitive) reasoning. Secondly, the way in which CONSYDERR handles evidence combination, weighted and cumulative is equated with a logic FEL; which in turn is a generalization of Horn clause logic and a modal logic. Thus CONSYDERR is well-equipped for capturing rule-based reasoning processes. Thirdly, representation in this architecture is closely tied to inferences to be performed (i.e., inferential structures are embedded within the representational structures and substrate constraints are incorporated into reasoning). Thus, a Hegelian *synthesis* of the *thesis* and the *antithesis* (i.e., rule-based reasoning and case-based reasoning) is achieved in high-level connectionist models.

The above comparison suggests some new directions for future research. The following questions need to be addressed, regarding high-level connectionist architectures:

- *Exploring architectural alternatives.* There are different ways of integrating various components of a hybrid system, and a better understanding of options and alternatives and their associated benefits and shortcomings is necessary for the further development of high-level connectionist architectures for addressing commonsense reasoning.
- *More sophisticated encoding of complex structures.* Currently only relatively simple structures are allowed in connectionist models, including high-level connectionist architectures. In order to deal with more complex types of commonsense reasoning, better encoding of complex structures is necessary.
- *Temporal reasoning.* Current connectionist models, including high-level connectionist architecture, can only deal with instantaneous information. Temporal and dynamic reasoning capabilities are necessary for handling some commonsense reasoning domains, including causal reasoning, and thus must be addressed. Recurrent structures used in some connectionist models might be utilized for dealing with the temporal dimension of commonsense reasoning, but a lot more work has to be done.

The continuous fusion of ideas from various AI paradigms, including rule-based reasoning, case-based reasoning, and connectionist models, will enable us to develop more complex, more robust intelligent systems for commonsense reasoning. In the process, a better understanding of human cognition can also be reasonably expected to be achieved.

Acknowledgments

This chapter is largely based on a paper published originally in *Fuzzy Sets and Systems*, 1995.

Chapter 6

Natural Language Processing with Subsymbolic Neural Networks

Risto Miikkulainen
University of Texas at Austin, USA
risto@cs.utexas.edu

6.1 Introduction

Natural language processing appears on the surface to be a strongly symbolic activity. Words are symbols that stand for objects and concepts in the real world, and they are put together into sentences that obey well specified grammar rules. It is no surprise that for several decades natural language processing research has been dominated by the symbolic approach. Linguists have focused on describing language systems based on versions of the universal grammar. Artificial intelligence researchers have built large programs where linguistic and world knowledge is expressed in symbolic structures, usually in LISP. Relatively little attention has been paid to various cognitive effects in language processing. Human language users perform differently from their linguistic competence, that is, from their knowledge of how to communicate correctly using language. Some linguistic structures (such as deep embeddings) are harder to deal with than others. People make mistakes when they speak, but fortunately it is not that hard to understand language that is ungrammatical or cluttered with errors. Linguistic and symbolic artificial intelligence theories have little to say about where such effects come from. Yet if one wants to build machines that would communicate naturally with people, it is important to understand and model cognitive effects in natural language processing.

The subsymbolic neural network approach holds a lot of promise for modeling the cognitive foundations of language processing. Instead of symbols, the approach is based on distributed representations that represent statistical regularities in language. Many cognitive effects arise naturally from such representations. In this chapter, the subsymbolic approach is first contrasted with the symbolic approach. Properties of distributed representations are illustrated, and examples of cognitive effects that arise are given. The achievements of the approach, in terms of subsymbolic systems that have been built so far, are reviewed and some remaining research issues outlined.

6.2   Subsymbolic representations

Symbolic and subsymbolic natural language processing systems are based on different strategies for representing information. In this section, the symbolic strategy is first contrasted with the subsymbolic (also called distributed) approach. Properties of distributed representations are then illustrated in a basic sentence case-role assignment task.

6.2.1   Properties of subsymbolic representations

Figure 6.1 shows a typical symbolic representation for an object, in this case car-32, expressed in LISP syntax. Such a representation could reside in the memory of a symbolic natural language processing (NLP) program, as part of its world knowledge. Several observations can be made on this representation. First, the symbols are discrete and disjoint. The symbol is either there or not, and if it is there, it is the symbol exactly. In other words, it is not possible for a symbol to exist at, e.g., 50% strength, or 90% accuracy. Second, the symbolic structure is grammatical. There are rules on how to read it: the first element in the list is the name of the whole list, and the following elements are slot–filler pairs that describe the object. Third, the structure is compositional and concatenative. It is possible to change the representation by just adding, changing, or deleting parts of it, and such changes do not affect any other part of the representation. For example, the owner might have been specified earlier as just *John*, but that symbol was replaced by a whole structure describing John. Such a change had no effect on any other part of the representation of car-32, or representations of any other objects in memory.

Figure 6.2 shows how such concepts could be represented subsymbolically in a neural network. Each concept is a *different pattern of activity* over the same set of units, and the processing knowledge about these concepts is superimposed on the same set of connection weights. A single unit does not represent any particular item, nor does an individual connection weight stand for any particular relation. Each unit and connection is involved in representing many different pieces of information. They stand for entities of finer granularity than symbols; therefore, representation and processing is subsymbolic.[1]

---

[1] Subsymbolic representations are also often called distributed representations, and each unit a microfeature. This is to contrast them with neural network implementations of symbolic representations, where each unit represents a separate item, or a separate semantic feature.

```
(car-32
    (owner (person-22
                (name John)
                (age 42)
                (married T)))
    (color red)
    (type sports)
    (age 5))
```

Figure 6.1. Symbolic representation of a concept: the representation consists of discrete and disjoint symbols, organized in a list structure that is grammatical and concatenative.

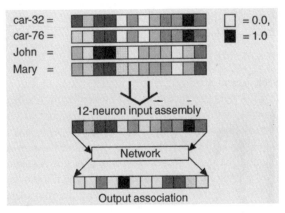

Figure 6.2. Subsymbolic representation: each item is represented as a pattern of activity over the same set of units, and the information about how to process the patterns is superimposed on the same set of connection weights. The patterns are continuous and can be more or less accurate. Similar concepts are represented by similar patterns, and the pattern as a whole is meaningful, not the values of individual units. The values are indicated by gray-scale coding in the figure. The output association could represent, for example, properties associated with the concept, such as name, age, type, and color.

Subsymbolic (i.e., distributed) representations have properties that are very different from the symbolic representations:

(i)   The representations are continuously valued.
(ii) Similar concepts have similar representations.
(iii) The representations are holographic, that is, any part of the representation can be used to reconstruct the whole.
(iv) Several different pieces of knowledge are superimposed on the same finite hardware.

The holographic property makes the system robust against noise, damage, and incomplete information. Because the same information is represented in several places, the processing is effectively based on an average of several representations. Noise is automatically filtered out in the averaging process, and loss of a few processing elements does not affect the average very much. The system does not selectively lose discrete blocks of information; instead, the accuracy of the output gradually degrades. Even when the input pattern is incomplete, the system can use the rest of the pattern to reconstruct the missing information. This property can be used in high-level systems to automatically create defaults and expectations.

The fourth property results in spontaneous generalization. This is the most important and most striking difference from the symbolic approach. It is not possible to change the representation encoded in a subsymbolic network without affecting the representations of all other items. Knowledge about an item is automatically generalized to all other items, to the degree that they are similar to that item. For example, when a new fact is learned about car-32, it is coded into the network by changing the connection weights. When the pattern for car-32 is input, the output of the network now shows the new fact. Because the weight changes are small and distributed over the entire set of weights, the output for *John* remains largely unchanged. This is because the input pattern for car-32 is very different from the pattern for John. The individual changes made in encoding the new fact about car-32 have mostly a random effect on *John*, canceling out on the average. However, the pattern for car-76 is very similar to *car-32*. When *car-76* is input to the network, the weight changes correlate with the input pattern very well, and the output for *car-76* now also shows the new fact. In other words, the new fact about car-32 is automatically generalized to car-76, to the degree that the representation for car-76 is similar to that for car-32.

6.2.2   Example: subsymbolic representations in sentence case-role assignment

Reading an input sentence into an internal representation is a most fundamental task in natural language processing. Below, properties of the subsymbolic representations will be illustrated in this task. The internal representation is based on the theory of thematic case roles [3]. The task consists of reading in the sentence word by word, and deciding which words play the roles of the act, agent, patient, instrument, and patient-modifier in the sentence.

For example, in *The ball hit the girl with the dog*, the subject *ball* is the instrument of the *hit* act, the object *girl* is the patient, the with-clause *dog* is a modifier of the patient, and the agent of the act is unknown. Role assignment is context-dependent: in *The ball moved*, the same subject *ball* is the patient. Assignment also depends on the semantic properties of the word. In *The man ate the pasta with cheese*, the with-clause modifies the patient; but in *The man ate the pasta with a fork*, the with-clause is the instrument. In yet other cases,

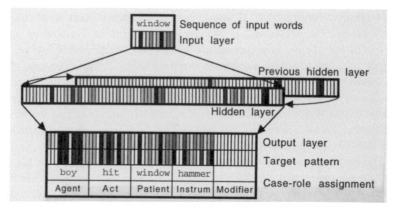

Figure 6.3. Subsymbolic case-role assignment of simple sentences. The network is a simple recurrent network that modifies word representations at its input layer as part of the learning process. In this snapshot, the network is in the middle of reading *The boy hit the window with the hammer.*

the assignment must remain ambiguous. In *The boy hit the girl with the ball*, there is no way of telling whether *ball* is an instrument of *hit* or a modifier of *girl*.

The architecture of the case-role assignment network is shown in figure 6.3. This network has one input assembly, for the current word in the input sequence, and five assemblies in the output, one for each case role. Through the backpropagation learning algorithm it is trained to map a sequence of input word representations into a stationary case-role representation. The network structure is based on Elman's [167] simple recurrent network (SRN). A copy of the hidden layer at time step t is saved and used along with the actual input at step $t + 1$ as input to the hidden layer. The previous hidden layer serves as a sequence memory, keeping track of where in the sequence the parser currently is and what has occurred before. The error is backpropagated and weights are changed at each step.

At the same time as the network learns the case-role assignment task, it develops distributed representations for the words. This is done through a method called FGREP [250]. Initially the representations for all words are assigned to random patterns. The representations are stored in a lexicon outside the network, and input and target patterns are formed by concatenating word representations currently in the lexicon. During each training presentation, the backpropagation error signal is propagated to the input layer, and the representations are changed as if they were an extra layer of weights. In other words, the network tries to modify the representations so that it could better perform the case-role assignment task. As a result, the final representations effectively code properties of the input words that are most crucial to the task.

The FGREP method allows the network to develop its own distributed input/output representations, a task which would be difficult to do by hand. However, it is important to note that backpropagation neural networks in general always develop distributed representations in their hidden layers as part of the learning. Such representations are learned essentially in the same process as the FGREP representations, and therefore the analysis below illustrates the properties of internal network representations in general.

The network was trained with data designed by [168] for the case-role assignment task. There were 19 sentence templates of the type *The human ate the food*. The actual sentences were formed by replacing the category words *human* and *food* with the actual members of the category, such as *man, woman, boy, girl*, and *chicken, cheese, pasta, carrot*. There were a total of 1475 sentences, with 30 different words. The network of course did not know about the templates and the categories, it only saw the actual sentences during training. To do the case-role assignment correctly, it had to learn the regularities in the examples and code them into the network weights and word representations.

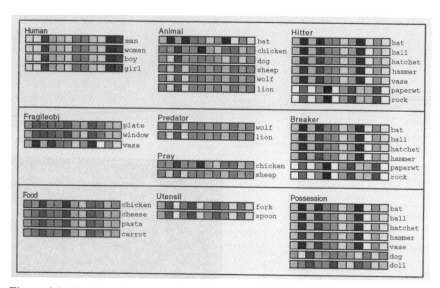

Figure 6.4. Distributed representations developed in the case-role assignment task. The representations are grouped according to categories used in generating the sentences. The representations were initially random. The network learned that certain words are used in similar ways in the data, and coded the similarities in the representations.

The network successfully learned the case-role assignment task, and the representations converged to a set where words that are used the same way in the data have similar representations (see example set in figure 6.4). One way to visualize the similarities among the 12-dimensional vectors is through a hierarchical merge clustering algorithm (figure 6.5). This process starts from the set of representation vectors. At each step, two representations closest to each other are found and merged into a single representation. In the process, clusters of similar representations become visible. The clusters found this way turn out quite similar to the noun categories that were used in generating the sentences. There are six prominent clusters with very small distances between items: animals, humans, utensils, two different types of hitter, and a combination of foods and fragile objects. Ambiguous words (*chicken*, which is an animal and food, and *bat*, an animal and hitter) and words with an unusual use (e.g., *dog*, which is a possession but not a hitter) do not fit very well into any category, and they are merged later in the process. The distances between the six clusters are quite large, indicating that they are well separated and spread out in the representation space. The cluster analysis demonstrates that the system develops subsymbolic representations that stand for the meanings of the words. Words that belong to similar categories have similar representations.

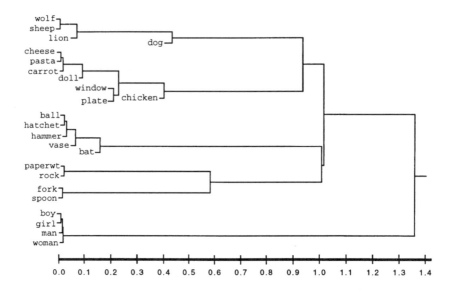

Figure 6.5. Visualizing the similarities in the representations through merge clustering. Step by step, the clusters with the shortest single linkage distance were merged. Distance is shown on the horizontal axis. The clustering follows the categories used in generating the sentences, showing that similar words have similar representations.

Inspection of the representations in figure 6.4 suggests that a single component does not play a crucial role in the classification of items. The fact that a word belongs to a certain category is indicated by the activity profile as a whole, instead of particular units being on or off. In other words, information is distributed over the entire representation vector in a holographic fashion. This is verified by inspection. The whole categorization is visible even in the values of a single component (figure 6.6). Each component provides a slightly

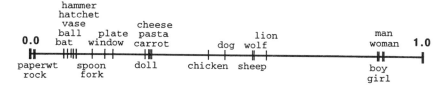

Figure 6.6. Holographic representations. The words are placed on a continuous line [0, 1] according to the value of the last component in their representations. The categorization is already evident in this single component.

different perspective on the data, and combining the values of more components provides an even more descriptive categorization. The complete representation can therefore be seen as a combination of 12 slightly different viewpoints into the word space. Such representations are very robust against errors. For example, one can eliminate units from the representation (by fixing them at 0.5 activation), and the decline in output accuracy is very gradual and approximately linear. Eliminating a unit means removing one classification perspective, and these perspectives apparently are additive.

Is it possible to name the properties the individual components are coding? The analysis of FGREP representations above suggests that the microfeatures in internal representations are identifiable only accidentally. Each individual component becomes sensitive to a combination of several features that is very unlikely to exactly match an established term, although partial matches are possible. For example, component 11 seems to separate animate objects, but the separation is not clear-cut and has a lot of finer structure (figure 6.6). Moreover, the network does not lose the information about animateness if this component becomes defective, so it would be incorrect to say that the component is responsible for representing 'animateness' of the input data.

The case-role assignment network generalizes very well to new sentences. If an unfamiliar sentence is meaningful at all, its representation pattern is necessarily close to something the network has already seen. This is because FGREP develops similar representations for similarly behaving words. For example, the network has never seen *The man ate the chicken with a fork*, but its representation is very close to the familiar sentence, *The girl ate the pasta with a spoon*, because the representation for *girl* is equivalent to *man*, and *fork* to *spoon*, and *chicken* is very much like *pasta*. In more general terms, the system can process the word x in situation S, because it knows how to process the word y in situation S, and the words x and y are used similarly in a large number of other situations.

The representations code not only properties of the words themselves, but also the contexts in which they occur. As a result, the network generates expectations and guesses automatically. This is demonstrated in figure 6.3. During training, the network is required to produce the complete output pattern after each step in the sequence, even before it is possible to unambiguously recognize the sequence. As a result, the output patterns at each step are averages of all possible event sequences at that point, weighted by how often they have occurred. After the next word is input, some of the ambiguities are resolved and correct patterns are formed in the corresponding assemblies. Often the sentence representation is almost complete before the sentence has been fully input. In figure 6.3, the network has read *The boy hit the window* and has unambiguously assigned these words to the agent, act, and patient roles. The instrument and modifier slots indicate expectations. At this point it is already clear that the modifier slot is going to be blank, because only human patients have modifiers in the data. Most probably, an instrument is going to be mentioned later, and an expectation for a hitter (an average of all possible hitters) is displayed in the instrument slot. If *with* is read next, a hitter is certain to follow, making the pattern even stronger. Reading a period next would instead clear the expectation in the instrument slot, indicating that the sentence is complete without this constituent. The expectations emerge automatically and cumulatively from the input word representations. Similarly to human language processing, the network can automatically fill in missing information, or select the correct sense for an ambiguous input word or guess meanings of unfamiliar words.

The properties discussed above make subsymbolic representations very different from symbolic representations. Philosophically, the two approaches are incompatible. Because of the fundamentally different way of representing information, it is not possible to *exactly* duplicate the function of a symbolic system with the function of a subsymbolic system. A representation cannot be distributed and symbolic at the same time [325]. Consequently, subsymbolic representations are not just a low-level implementation of symbolic representations, but a fundamentally different approach to modeling natural language processing.

6.3 Modeling human language processing

People seem to have two fundamentally different mechanisms at their disposal for performing cognitive tasks. Following a sequential symbolic strategy is the more obvious of the two. One does not have an immediate answer to the problem, but the answer is sequentially constructed from stored knowledge by a high-level goal-directed process, that is, by reasoning. Another type of cognitive processing occurs through associations immediately, in parallel, and without conscious control, in other words, by intuition. Large amounts of information, which may be incomplete or even conflicting, are simultaneously brought together to produce the most likely answer.

For cognitive processes based on conscious rule application, symbolic systems are a good approximation. However, intuitive processing cannot be easily implemented symbolically. In contrast, neural networks represent knowledge in terms of statistical regularities in their training, and processing is opaque, nonconcatenative, and immediate. Therefore, neural networks fit very well into modeling intuitive inference. A major issue is: are humans indeed symbol processors at the high level, for example in processing language, or could such processes be an epiphenomenon of low-level associative and statistical mechanisms?

6.3.1   Symbols versus soft constraints

Processing sentences with embedded clauses is one task that shows how human language processing, although clearly symbolic at the surface level, at closer look exhibits strong subsymbolic qualities. Consider the following sentence:

The girl who the boy hit cried.

This sentence has a relative clause attached to the main noun *boy*. Relative clauses have a simple structure, and it is easy to form deeper embeddings by repeating the structure recursively:

The girl who the boy who the girl who lived next door blamed hit cried.

This sentence contains no new grammatical constructs. The familiar embedded clause construct is used just three times, and the resulting sentence is almost incomprehensible. If humans were truly symbol processors, the number of levels would make no difference. It should be possible to handle each new embedding just like the previous one.

Now consider a similar sentence:

The car that the man who the dog that had rabies bit drives is in the garage.

This sentence has the same grammatical structure as the previous one, and for a symbol processor it should be equally easy, or hard, to process. Yet somehow this sentence is understandable, whereas the previous one was not. What is different?

Whereas the second sentence could be understood only based on syntactic analysis, the third one has strong semantic constraints between constituents. We know that dogs have rabies, people drive cars, and cars are in garages. These constraints make it possible for a human to understand the sentence even when they lose track of its syntactic structure. A symbol processor can parse the syntactic structure of both sentences perfectly well, and receives no benefit from the semantic constraints.

The conclusion from these examples is that people are not pure symbol processors when they understand language. Instead, all constraints—grammatical, semantic, discourse, pragmatic—are simultaneously taken into account to form the most likely interpretation of the sentence. This behavior is difficult to model with symbolic systems, but it is exactly what neural networks are good at. The example discussed in the next section illustrates how.

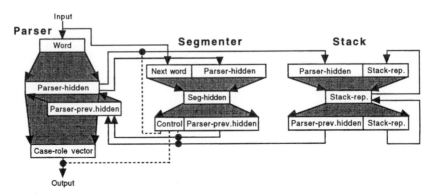

Figure 6.7. Subsymbolic case-role assignment of sentences with embedded clauses. The model, SPEC, consists of the parser (a simple recurrent network), the stack (a RAAM network), and the segmenter (a feedforward network). The gray areas indicate propagation through weights, the solid lines stand for pattern transport, and the dashed lines represent control outputs (with gates).

6.3.2   Example: case-role assignment of embedded clauses

The SPEC system (figure 6.7; [251]) is an extension of the subsymbolic parser architecture of section 6.2.2, designed to deal with sentences with embedded clauses. SPEC receives a sequence of word representations as its input, and for each clause in the sentence, forms an output representation indicating the assignment of words into case roles. The case-role representations are read off the system and placed in a short-term memory (currently outside SPEC) as soon as they are complete. SPEC consists of three main components: the parser, the segmenter, and the stack. The parser is a network similar to the parser above. The case-role assignment is represented at the output of the parser as a case-role vector (CRV), that is, a concatenation of those three word representation vectors that fill the roles of agent, act, and patient in the sentence. For example, the word sequence *The girl saw the boy* receives the case-role assignment agent = *girl*, act = *saw*, patient = *boy*, which is represented as the vector |*girl saw boy*| at the output of the parser network. When the sentence consists of multiple clauses, the relative pronouns are replaced by their referents: *The girl who liked the dog saw the boy* parses into two CRVs: |*girl liked dog*| and |*girl saw boy*|.

The parser receives a continuous sequence of input word representations as its input, and its target pattern changes at each clause boundary. For example, in reading *The girl who liked the dog saw the boy* the target pattern representing |*girl saw boy*| is maintained during the first two words, then switched to |*girl liked dog*| during reading the embedded clause, and then back to | *girl saw boy*| for the rest of the sentence. The CRV for the embedded clause is read off the network after *dog* has been input, and the CRV for the main clause after the entire sentence has been read. When trained this way, the network is not limited to a fixed number of clauses by its output representation. Also, it does not have to maintain information about the entire past input sequence in its memory, making it possible in principle to generalize to new clause structures. Unfortunately, after a center embedding has been processed, it is difficult for the network to remember earlier constituents. This is why a stack network is needed in SPEC.

The hidden layer of a simple recurrent network forms a distributed representation of the sequence so far. The stack has the task of storing this representation at each center embedding, and restoring it upon return from the embedding. For example, in parsing *The girl who liked the dog saw the boy*, the hidden-layer representation is pushed onto the stack after *The girl*, and popped back to the parser's previous-hidden-layer assembly after *who liked the dog*. In effect, the SRN can then parse the top-level clause as if the center embedding had not been there at all.

The stack is implemented as a recursive auto-associative memory (RAAM; [19]). RAAM is a three-layer backpropagation network trained to perform an identity mapping from input to output. As a side effect, the hidden layer learns to form compressed distributed representations of the network's input/output patterns, which consist of the top element of the stack and the distributed representation of the rest of the stack. The hidden layer representation is then recursively used as the representation for the rest of the stack in the next push operation. A potentially infinite stack can be compressed into a fixed-size representation this way. The stack structure can later be decoded by loading its distributed representation into the hidden layer and reading off the top element and the distributed representation for the rest of the stack at the output.

The parser + stack architecture alone is not quite sufficient for generalization into novel relative clause structures. For example, when trained with only examples of center embeddings (such as the above) and tail embeddings (like *The girl saw the boy who chased the cat*), the architecture generalizes well to new sentences such as *The girl who liked the dog saw the boy who chased the cat*. However, the system still fails to generalize to sentences like *The girl saw the boy who the dog who chased the cat bit*. Even though the stack takes care of restoring the earlier state of the parse, the parser has to learn all the different transitions into relative clauses. If it has encountered center embeddings only at the beginning of the sentence, it cannot generalize to a center embedding that occurs after an entire full clause has already been read.

This problem can be overcome with the segmenter network. The segmenter is trained to recognize the transition to the relative clause, and to modify the hidden-layer pattern so that only the relevant information remains (i.e., *boy* in the above example). In other words, the controller has an internal representation for the 'relative clause' construct, and applies it to change the hidden-layer pattern so that the low-level sequence-processing network only has to deal with one type of clause boundary. By dividing the task into meta-level control, low-level pattern transformation, and memory, the whole system can generalize to novel structures.

SPEC was trained with 100 examples each of two sentence structures: (1) the two-level tail embedding (such as *The girl saw the boy who chased the cat who the dog bit*) and the two-level center embedding (e.g., *the girl who the dog who chased the cat bit saw the boy*), and the stack was trained to store up to three levels of center embeddings. After training, SPEC generalized perfectly to all other combinations of center and tail embeddings of four clauses, that is, to 98 100 different sentences.

The main result, therefore, is that the SPEC architecture successfully generalizes not only to new instances of the familiar sentence structures, but to new structures as well. However, SPEC is not a mere reimplementation of a symbol processor. As SPEC's stack becomes increasingly loaded, its output becomes less and less accurate; symbolic systems do not have any such inherent memory degradation. An important question is, does SPEC's performance degrade in a cognitively plausible manner, that is, does the system have similar difficulties in processing center embeddings as people do?

To elicit enough errors from SPEC to analyze its limitations, the Stack's performance was degraded by adding 30% noise in its propagation. Such an experiment can be claimed to simulate overload, stress, cognitive impairment, or lack of concentration situations. The system turned out to be remarkably robust against noise. The average parser error rose somewhat, but the system still got 94% of its output words right. As expected, most of the errors occurred as a direct result of popping back from center embeddings with an inaccurate previous-hidden-layer representation. For example, in parsing *The girl, who the dog, who the boy, who chased the cat, liked, bit, saw the boy*, SPEC had trouble remembering the agents of *liked*, *bit* and *saw*, and patients of *liked* and *bit*. The performance depends on the level of the embedding in an interesting manner. It is harder for the network to remember the earlier constituents of shallower clauses than those of deeper clauses (figure 6.8). For example, SPEC could usually connect *boy* with *liked* (in 80% of the cases), but it was harder for it to remember that it was the *dog* who *bit* (58%) and even harder the *girl* who *saw* (38%) in the above example.

Such behavior seems plausible in terms of human performance. Sentences with deep center embeddings are harder for people to remember than shallow ones [150]. It is easier to remember a constituent that occurred just recently in the sentence than one that occurred several embeddings ago. Interestingly, even though SPEC was especially designed to overcome such memory effects in the parser's sequence memory, the same effect is generated by the stack architecture. The latest embedding has noise added to it only once, whereas the earlier elements in the stack have been degraded multiple times. Therefore, the accuracy is a function of the number of pop operations instead of a function of the absolute level of the embedding.

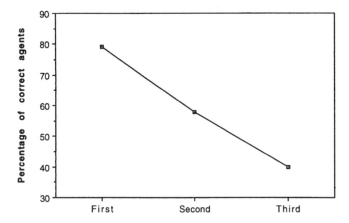

Figure 6.8. Memory accuracy after return from center embeddings. The percentage of correctly remembered agents is plotted after the first, second, and the third pop in sentences with three levels of center embeddings. Each successive pop is harder and harder to do correctly. Similarly, SPEC remembers about 84% of the patients correctly after the first pop, and 67% after the second pop. The stack representations were degraded with 30% noise to elicit the errors.

When the SPEC output is analyzed word by word, several other interesting effects are revealed. Virtually in every case where SPEC made an error in popping an earlier agent or patient from the stack it confused it with another noun (54 556 times out of 54 603; random choice would yield 13 650). In other words, SPEC performs plausible role bindings: even if the exact agent or patient is obscured in the memory, it 'knows' that it has to be a noun. Moreover, SPEC does not generate the noun at random. Out of all nouns it output incorrectly, 75% had occurred earlier in the sentence, whereas a random choice would give only 54%. It seems that traces for the earlier nouns are discernible in the previous-hidden-layer pattern, and consequently, they are slightly favored at the output. Such priming effect is rather surprising, but it is very plausible in terms of human performance.

To test the effects of semantic constraints on sentence processing performance, the training sentences had been generated with a number of restrictions. A verb could have only certain nouns as its agent or patient. Some verbs had only one possible agent or patient, others had two, three or four. This way semantic restrictions of different strengths could be introduced in the training data. The restrictions turned out to have a marked effect on the performance (figure 6.9). If the agent or patient that needs to be popped from the stack was strongly correlated with the verb, it is easier for the network to remember it correctly. The effect depends on the strength of the semantic coupling. For example, *girl* is easier to remember in *The girl, who the dog bit,*

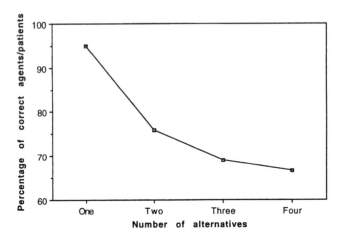

Figure 6.9. Effect of the semantic restrictions on the memory accuracy. The percentage of correctly remembered agents and patients over the entire corpus is plotted against how strongly they were semantically associated with the verb. When there was only one alternative (such as *dog* as an agent for *bit* or *cat* as the patient of *chased*), SPEC remembered 95% of them correctly. There was a marked drop in accuracy with two, three and four alternatives. The stack representations were degraded with 30% noise to elicit the errors.

liked the boy, than in *The girl, who the dog bit, saw the boy*, which is in turn easier than *The girl, who the dog bit, chased the cat*. The reason is that there are only two possible agents for *liked*, whereas there are three for *saw* and four for *chased*. While SPEC gets 95% of the unique agents right, it gets 76% of those with two alternatives, 69% of those with three, and only 67% of those with four.

A similar effect has been observed in human processing of relative clause structures. Half the subjects in Stolz's [265] study could not decode complex center embeddings without semantic constraints. [264] showed that young children understand embedded clauses better when the constituents are semantically strongly coupled, and [2] observed similar behavior in aphasics. This effect is often attributed to limited capability for processing syntax. The SPEC experiments indicate that it could be at least partly due to impaired memory as well. When the memory representation is impaired with noise, the parser has to clean it up. In propagation through the parser's weights, noise that does not coincide with the known alternatives cancels out. Apparently, when the verb is strongly correlated with some of the alternatives, more of the noise appears coincidental and is filtered out.

The conclusion from the SPEC system is, then, that even when the subsymbolic architecture is designed for strong linguistic performance, such as generalizing to novel relative clause structures, it can exhibit subsymbolic effects similar to those of humans. Deep center embeddings are difficult for SPEC, and semantic constraints make the task easier. It is by tapping into such phenomena that the subsymbolic models can be most useful in natural language processing.

6.4   Overview of subsymbolic natural language processing

Natural language processing has been an active area of connectionist research for over a decade. Subsymbolic models have been developed to address a variety of issues, such as semantic interpretation, learning syntax and semantics, prepositional phrase attachment, anaphora resolution, morphology, phrase generation, identification of discourse topics, and goal-plan analysis [123, 248, 223, 266].

A good amount of work has been done showing that networks can capture grammatical structure. For example, [79] showed how simple recurrent networks can learn a finite-state grammar. These networks are similar to those discussed above, except they are trained to predict the next item in the sequence instead of reading the sequence into a stationary representation. [79] trained such an SRN with sample strings from a particular grammar, and it learned to indicate the possible next elements in the sequence. For example, given a sequence of distributed representations for elements B, T, X, X, V, and V, the network turns on two units representing X and S at its localist output layer, indicating that in this grammar, the string can continue with either X or S.

[167] used the same network architecture to predict a context-free language with embedded clauses. The network could not learn the language completely, but its performance was remarkably similar to human performance. It learned better when it was trained incrementally, first with simple sentences and gradually including more and more complex examples. The network could maintain contingencies over embeddings if the number of intervening elements was small. However, deep center embeddings were difficult for the network, as they are for humans. [151] further showed that center embeddings were harder for this network than right-branching structures, and that processing was aided by semantic constraints between the lexical items. Such behavior matches human performance very well.

The above architectures demonstrated that subsymbolic networks build meaningful internal representations when exposed to examples of strings in a language. They did not address how such capabilities could be put to use in parsing and understanding language. [168] first identified the sentence case-role assignment as a good approach. The approach is particularly well suited for neural networks because the cases can be conveniently represented as assemblies of units that hold distributed representations, and the parsing task becomes that of mapping between distributed representation patterns. McClelland and Kawamoto showed that given the syntactic role assignment of the sentence as the input, the network could assign the correct case roles for each constituent.

The network also automatically performed semantic enrichment on the word representations (which were hand-coded concatenations of binary semantic features), and disambiguated between the different senses of ambiguous words.

As was discussed above in section 6.2.2, essentially the same task can be performed from sequential word-by-word input by a simple recurrent network and meaningful distributed representations for the words can be automatically developed at the same time. Systems with FGREP representations generally have a strong representation of context, which results in good generalization properties, robustness against noise and damage, and automatic 'filling in' of missing information. The FGREP representations can be augmented with ID information, which allows the system to process a large vocabulary even after learning only a small number of distinct meanings. In this ID + content approach, representations for, e.g., *John*, *Bill*, and *Mary* are created from the FGREP representation of *human* by concatenating unique ID patterns in front of it. All these words have the same meaning for the system, and it knows how to process them even if it has never seen them before.

St John and McClelland [98] further explored the subsymbolic approach to sentence interpretation in a 'sentence gestalt' model. They aimed at explaining how syntactic, semantic, and thematic constraints are combined in sentence comprehension, and how this knowledge can be coded into the network by training it with queries. The gestalt is a hidden-layer representation of the whole sentence, built gradually from a sequence of input words by a simple recurrent network. The second part of the system (a three-layer backpropagation network) is trained to answer questions about the sentence gestalt, and in the process, useful thematic knowledge can be injected into the system. Similar approach can also be applied to processing script-based stories [97].

A number of researchers have proposed modular and structured architectures. In addition to SPEC, FGREP networks were used to build a story processing system called DISCERN [250]. DISCERN is a large-scale natural language processing system implemented entirely at the subsymbolic level. DISCERN aims at bridging the gap between subsymbolic mechanisms and complex high-level behavior. Subsymbolic neural network models of parsing, generating, reasoning, lexical processing, and episodic memory are integrated into a single system that learns to read, paraphrase, and answer questions about stereotypical narratives. In this approach, connectionist networks are not only plausible models of isolated cognitive phenomena, but also serve as building blocks for large-scale artificial intelligence systems.

In Jain's [202] structured incremental parser, one module was trained to assign words into phrases, and another to assign phrases into case roles. These modules were then replicated multiple times so that the recognition of each constituent was guaranteed independent of its position in the sentence. In the final system, words were input one at a time, and the output consisted of local representations for the possible assignments of words into phrases, phrases into clauses, phrases into roles in each clause, and for the possible relationships of the clauses. A consistent activation of the output units represented the interpretation of the sentence. The system could interpret complicated sentence structures, and even ungrammatical and incomplete input. The parse result was a description of the semantic relations of the constituents; the constituents themselves were not represented.

Berg's [115] XERIC and Sharkey and Sharkey's [95] parser were both based on the idea of combining a simple recurrent network with a recursive auto-associative memory (RAAM) that encodes and decodes parse trees. In Sharkey and Sharkey's model, firstly the RAAM network was trained to form compressed representations of syntactic parse trees. Secondly, an SRN network was trained to predict the next word in the sequence of words that make up the sentence. Thirdly, a standard three-layer feedforward network was trained to map the SRN hidden-layer patterns into the RAAM parse-tree representations. During performance, a sequence of words was first read into the SRN, its final hidden layer transformed into a RAAM hidden layer, and then decoded into a parse tree with the RAAM network. Berg's XERIC worked in a similar manner, except the SRN hidden-layer representations were directly decoded by the RAAM network.

Capabilities of RAAM networks and the distributed representations they form have been extensively studied (see [41, 59]). Although the constituents of such representations, e.g., words in the parse tree, are not directly available, it is possible to perform 'holistic' transformations on the entire patterns. For example, [63] trained one RAAM network to encode sentences in active voice, such as *John loves Michael*, and another RAAM network to encode same sentences in the passive, such as *Michael is loved by John*. A third, feedforward network was trained to map the distributed representation of the active sentence to the representation of the passive. The transformation network easily generalized to new sentences, showing that it had developed a sensitivity to structure that was only implicitly encoded in the distributed representations. Chrisman [67] applied the same idea of holistic transformations to translating between English and Spanish sentences. Instead of two RAAMs and a transformation network he used two RAAMs with a common hidden layer. This forced the English and Spanish sentences to be encoded with more similar distributed representations, resulting in better performance.

The above results indicate that subsymbolic networks can represent and process linguistic knowledge in a cognitively valid manner, with strong sensitivity to context. They also show promise that complex structures can be processed, and large systems can be built from subsymbolic components.

6.5 Future challenges

Even though the above systems are successful in what they are designed to do, and show very interesting cognitive behavior, they are still mostly demonstrations of capabilities on toy problems. Before subsymbolic natural language processing systems will rival the large symbolic NLP systems, several issues must be resolved:

- How can complex linguistic representations be encoded on neural networks? For example, how can you represent an indefinite number of agents in a clause, or clauses in a sentence, or sentences in a story, when you only have a limited and fixed number of units in the network? People do not seem to have a fixed upper bound, although there clearly are memory limits. It is possible that some kind of reduced description, similar to those modeled by RAAM, is being formed. However, so far it has turned out very difficult to make the RAAM architecture generalize to new structures. For example in Sharkey and Sharkey's parser, and in Chalmers's and Chrisman's transformation systems, the limiting factor was the RAAM representation, not the transformation.
- How can we come up with training examples for realistic language processing? Although large corpora of unprocessed text are readily available, subsymbolic systems usually require more sophisticated information as targets, such as case-role representations for parsing, or transfer examples for translation. Building such corpora is very laborious, and it is unclear whether it is ever possible to have large enough training sets, as long as generalization is limited to interpolation between training examples.

Building systems that would be able to generalize in a more fundamental way, by dynamic inferencing, or bringing together processing knowledge that has previously been seen in separate situations, is perhaps the greatest challenge for connectionist systems. It is only possible by strongly constraining the kinds of thing the networks do, as in SPEC: the parser was limited to only simple pattern transformation, and the structure of the network forced generalization to novel structures. Time will tell how far such solutions will take us, but it is possible that network architectures and learning algorithms can be designed that would be able to learn metaknowledge about the kinds of tasks they are performing, and would allow them to use context when it is useful, and ignore it when it is irrelevant.

6.6 Conclusion

In this chapter, foundations for subsymbolic natural language processing were reviewed. Distributed representations were found to have very different properties from symbolic representations, properties that match the cognitive constraints in language processing very well. Two parsing architectures were discussed, one where properties of distributed representations could be easily illustrated, and another more complex architecture where the emergent cognitive effects could be analyzed. An overview of the state of the art in subsymbolic natural language processing systems suggests that many language phenomena can be modeled this way, although it is still difficult to scale the approach up to the level of complexity required by natural language processing in the real world.

Acknowledgments

This research was supported in part by the Texas Higher Education Coordinating Board under grant ARP-444.

Note

Source code for the FGREP, SPEC, and DISCERN systems, as well as an on-line demo of DISCERN, are available on the World Wide Web at URL http://www.cs.utexas.edu/users/nn.

Chapter 7

The Relational Mind

Professor John G Taylor
Kings College, UK
udah057@kcl.ac.uk

7.1 Introduction

Consciousness is presently at the centre of attention. With the demise of strict behaviourism has come a flood of activity attempting to understand this highest level of human activity. With the development of better understanding of the possible modes of action of neural networks, both at the artificial and living levels [310], it should be possible to explore models of the mind with a realistic relation to the actual physical structures involved. This would allow better cognizance to be taken of the increasing wealth of data obtained from recordings of dynamic brain activity over the vast range of aggregates, from single cells or cell groups obtained from intercellular electrodes up to areas, or even whole brains, by positron emission tomography (PET), electroencephalography (EEG), and magnetoencephalography (MEG) techniques.

We start with a discussion of the problems of consciousness. A description of a global model of the mind [307] is then given, and the question of the uniqueness of consciousness considered in terms of a specific neural structure in the brain. Experimental evidence in support of the resulting model for the uniqueness of consciousness is discussed [176]. The phenomenon of 'backward referral in time' that Libet *et al* discovered is described and its implications in terms of underlying neural structures explored. In terms of this the manner in which awareness might arise is then followed up in terms of subliminal effects. This leads to a more general view of consciousness, which is then applied to attempt to answer some of the hard questions on consciousness. Finally the manner in which the general model might be tested is explored in terms of the recent PSYCHE global brain modelling project.

7.2   Problems of consciousness

There are numerous problems which have been raised about the nature of consciousness, and many conjectures as to their solution. The problems can be divided into the 'soft' and 'hard' questions. The soft problems are concerned with the underlying neural processes which are required to support consciousness, and whose modifications by drugs, brain injury or disease can cause considerable changes to personality and effectiveness in response to the tasks of every-day life. The answers to these soft problems are difficult, but are gradually being obtained by the enormous modern programme of research exploring the brain at all levels, from single nerve cell up to global activity involved in attention or other psychological states.

On the other hand the hard problems are of a somewhat different character, and involve the inner nature of consciousness itself. There is a personal subjective character to consciousness, related to the 'raw feels' or so-called qualia, over which there has been considerable debate. There are also the features of subjectivity, intentionality, and introspection which are also apparent components of consciousness. The former of these is regarded as impossible to attack from the objective viewpoint of modern science [203], but we will attempt to give a response to this problem towards the end of this chapter.

Intentionality involves the question of the meaning which we are able to attach to our mental concepts; the problem of meaning and the construction of concepts are discussed elsewhere from an active viewpoint [312]. Finally introspection involves the nature of self, which will also be discussed later.

7.3   The relational mind model

The basic idea behind this approach is that consciousness arises due to the active comparison of ongoing brain activity, stemming from external inputs in the various modalities, with somewhat similar past activity stored in semantic and episodic memory. The mental content of an experience therefore contains, as a component, the set of relations of that experience to stored memories of relevant past experiences. Thus the consciousness of the blue of the sky, as seen now, is determined by the stored memories of one's past experience of blue skies, say on hillsides, at the seaside stretched flat on the sands, or in one's garden sitting in a chair. Not only episodic memory need be involved; there may also be semantic memory of objects of a visual or other scene. The semantic priming thereby involved may itself also lead to further episodic memory activation, in which further relations to past experience are accessed.

The mind, in this approach, may thus be summarized as being created by activation of a set of related memory states by the present input. This relational structure has been explored theoretically elsewhere [306] and more neurophysiologically [307]. It leads to the relational theory of meaning or intentionality, in which the activation of relevant past memories allows an object to be recognized in terms of its physical structure, actions that can be taken with respect to it, and so on. This approach to meaning is accepted by some neuropsychologists [61] and by most cognitive psychologists [146]. The inner, private nature of consciousness is thereby explicable in terms of the internal storage of past experiences, unique to the individual.

There are many questions that arise from this global approach. One of these is as to the nature of the control system [7] which leads to a seemingly unique stream of consciousness [308]. We will explore this from a new experimental angle in the next section.

7.4   Global control

One would expect there to be control structures in the brain which:

- Only allow certain memories to be activated, which are related to the corresponding input.
- Are involved with assessing the level of support to be given to any new incoming input from activated stored patterns in any competition that may then ensue between possible sources of awareness.

Both of these aspects have been considered [308] in terms of possible networks which could perform pattern matching. All input to the cortex is first sent to the thalamus in the middle of the brain, and then on to the cortex. In the process it has to pass through, and feed, activity to a thin sheet of inhibitory neurons, called the nucleus reticularis thalami (NRT), shown in figure 7.1. It was suggested that the NRT may allow global control of cortical activity to be achieved due to the extensive interconnection of the neurons with each other on the sheet. The combined activity of the NRT and cortex may support such competitive activity, with the NRT functioning as an inhibitory sheet similar in construction and function to the outer layer of neurons in the retina. This latter can be modelled as a network in which activity disperses in a well defined manner like heat diffusion in a metal bar [305]. However, since the NRT effectively has a negative diffusion coefficient (from its mutually inhibitory neurons) then it can sustain 'bunched' spatially inhomogeneous activity, in which the opposite to dispersion is occurring. Competition between neighbouring thalamic or cortical inputs can be seen to be occurring, since one patch of NRT activity tries to build itself up at the expense of activity elsewhere on the NRT sheet. In this manner the sheet may function as a global controller of cortical activity, allowing a global form of competition between several distinct inputs to be run.

Interesting results, now more than thirty years old [176], appear relevant to the nature of global control. The experiments of Libet were to determine the threshold current for conscious experience when a 1 mm diameter stimulating electrode was placed on the post-central gyrus and the just conscious experience of what seemed like a localized skin stimulus reported by the patient. The stimulus was delivered as a series of short (of around 0.1 ms duration) pulses.

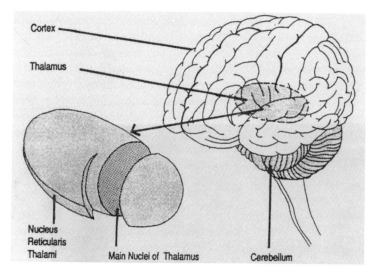

Figure 7.1. Position of the NRT in the brain.

There are two features of the data which stand out, which can be summarized as two quantitative laws:

- Law 1. For threshold current to be consciously experienced over a short (<0.5 s) duration, the applied electrical energy (frequency times duration times square of current) must be greater than a critical value.
- Law 2. For a duration longer than about 0.5 s the applied electrical power must be large enough to allow the conscious experience to continue.

The requirement of enough applied electrical energy to capture, or turn on, conscious awareness in the short term would seem to fit well with the NRT control structure model above. For that is functioning essentially as a resistive circuit, with some non-linearity to provide stability, and such a circuit would be expected to function in terms of the electrical energy requirements for capture of the dominant mode.

The second result above leads to a need for enough injected power to keep the control system going; there will be a certain amount dissipated, and so power above that critical level will have to be injected to hold the control of consciousness achieved by the earlier injected electrical energy. Simulations have shown that the features described above are indeed contained in the model based on the interaction of the NRT with the cortex [7].

It is possible that other approaches may be derived to explain the above results. Thus, they appear similar to many psychophysical results on threshold discrimination levels, such as Bloch's law in vision [34]. This law states that for visual input patterns with persistence less than some 100 ms the discrimination level satisfies a law similar to law 1, where electrical energy in that law is now replaced by light, primary cortical region by retina and electrode by light surface. For times beyond 100 ms, law 2 applies (again with the same replacements). The simplest explanation of this (and similar) laws is in terms of peripheral summation. Thus rods and cones have a temporal integration period of 200 and 30 ms respectively [34]. Such an explanation cannot be easily used in the case of direct cortical excitation, since the temporal summation time has been found to be about 50–100 ms [176]; this value is very different from the 500 ms or so value of the minimum duration for a pulse at liminal current intensity needed to cause the occurrence of conscious awareness.

7.5 Backward referral in time

The phenomenon of 'backward referral in time' has caused a great deal of discussion. There are clearly controversial questions of an experimental nature that may still need to be resolved about the phenomenon, but let us accept the available results of Dennett and Kinsbourne [86]. We can summarize them as to the nature of the stimulus needed to achieve conscious awareness (a continued train of pulses verses a single pulse) and the presence or not of backdating of the first experience of the stimulus. A peripheral stimulus to the skin only requires a single pulse, which, however, gets backdated, which also happens to a pulse train of 'neuronal adequacy' duration applied thalamically. The cortical stimulus neither gets backdated nor can be achieved by a single pulse, but only by a pulse stream of temporal duration equal to the neuronal adequacy time.

We wish to propose an explanation of these features which is consistent with, and supports, the competitive model of consciousness presented in the previous section. Let us suppose that the negative after-potentials noted in response to a single pulse skin stimulus [175] correspond to that input having rapidly gained access to the appropriate working memory, and continuing to be active there. That continuation of activity is, indeed, our definition of a working memory or WM, as we will denote it, and agrees with increasing numbers of observations of such continued activity in various brain sites under trial conditions in monkeys. The evidence indicates that more artificial sources of input from thalamus or cortex are not able to gain the direct access to the working memory that the single skin pulse did. Instead the former inputs have to be injected by creating their own working memory, in other words as a train of pulses, not a single pulse. The mechanism by which the skin pulse can activate the WM rapidly (in, say, 50 ms), whilst the other inputs cannot, may be in terms of what is called the mesencephalic reticular activating system (MRAS), a reticulum of nerve cells in the brainstem which is thought to sample all inputs coming to the cortex. The action of the MRAS is thought to be excitatory to the NRT, and as such may convey input directly to the appropriate part of NRT so that the skin input, processed through somato-sensory cortex, easily activates the appropriate working memory. This activity then proceeds to attempt to win the competition on the NRT, as discussed in the previous section.

We note that in the competitive model it was found by simulation that however strong an artificial input is to the cortex, it always takes a minimal but non-zero time to win the competition. This agrees with the reported results [175]. Moreover a strong single pulse to the thalamus was reported there as not causing awareness, due to neuronal adequacy of the input being needed to cause such awareness. The same happens in the simulations, when a large new input is rapidly switched off almost immediately after being turned on—there is little change in NRT activity (since it takes time to build up), and little chance to win the competition.

The backdating that was found to occur has a different pattern of distribution from that of the nature of the requisite stimulus (single pulse versus pulse train). A direct sub-cortical input, most likely at the thalamic level, appeared necessary to achieve backward referral in time. One conjecture as to how that might occur, consistent with the presence of WM structures, is that a thalamic-level timing mechanism is turned on by the peripheral or thalamic input on its first appearance. This is then updated constantly, essentially as a counter, until the neuronal adequacy is reached; when the counter is reset to zero. The contents of the counter are used to tag the input time and so allow for the backward temporal referral. If this occurs for two competing inputs, then the first counter has a larger or stronger output, so could give the conscious experience of having occurred before the second, later input. Such a timing mechanism is expected to be hard-wired, since there would appear to be little value in having it adaptive. However, the size of the timing interval may be modified by neuromodulatory effects, so leading to the well known variations of subjective time in, say, emergency situations.

A timer of the above sort has to be able to count up to about 500 ms, and do so with an accuracy of, say, 25 ms (although other units of time, such as 100 ms, would also do, related to 10 Hz oscillations, as compared to 40 Hz giving 25 ms). Thus it must be able to count at least 20 different states, each leading to the next as succeeding intervals of 25 ms pass. That could be achieved by a recurrent network with recurrence time equal to some fraction of 25 ms. Each recurrence would lead to recruitment of an identical number m of neurons, so that after time n there would be mn neurons active. Different inputs would recruit from different areas of the timing network, and would then compete against each other in their contributions to the input winning the consciousness competition. Undoubtedly other neuronal models of timing circuits exist, although the one proposed above appears to be one of the simplest. We note that this timed competitive approach to consciousness by WMs can also give an explanation of the color-phi and the cutaneous rabbit phenomena [86], although there is not space to enlarge on that here.

One of the clear predictions of the above discussions is the existence of timing circuitry, possibly in thalamic regions, although there could be other sites, such as in cerebellum or basal ganglia. This timing mechanism would be activated by a peripheral thalamic stimulus, and would be predicted to give an increasing input until neuronal adequacy occurred. Such a linear ramp function would have a clear signature, although it might only be seen, for example, by non-invasive techniques, such as by multi-channel MEG measurements, at the short time scales involved. A further prediction is that a direct cortical input would not be able to activate the timing circuitry, so not allowing backward referral in time. At the same time there are further predictions as to the activation of suitable WM circuitry by a peripheral stimulus, whilst the absence of such activity would be clear if thalamic or cortical stimulation were used. Measurement of MRAS activity during such tasks would be important here to probe the activity further.

7.6 Subliminal effects

The question which must now be attacked, from the position reached in the chapter, is as to the more detailed manner in which consciousness of a given input can arise from the host of competing inputs and the objects which they represent. It is clear that attention must ultimately be included in such an analysis, but attention has already been focused on the appropriate region of space by subjects in the experiments to be discussed shortly, so that initially only processing will be considered in which there are no distractors in the visual or other input space. In such conditions it is not expected that attention will play a crucial part in the processing.

Important work on the nature of non-conscious effects on later processing has already been done [185]. There is a host of further material on this topic, but only Marcel's work will be considered here in detail since it is most apposite.

The work under consideration [185] involves the processing of three consecutive letter strings, the first and third being responded to as to whether or not they are a word. The second word is viewed under one of three possible conditions: as visually perceptible (for half a second), as energy masked (so not perceived at all) or with a pattern mask after a 10 ms exposure, the pattern mask being transmitted soon enough after the brief exposure to the word so that there is no conscious experience of the word.

Six possible relations of associativity were chosen between the three words (when they were words), when the middle word had an ambiguous meaning, such as the word 'palm'. Thus in the so-called incongruent case the consecutive words might be 'hand', 'palm' and 'wrist'; in the unbiased case they could be 'clock', 'palm' and 'wrist'. The main question addressed was as to the manner in which conscious or subliminal processing of the second word, possibly biased by the first word, might affect the reaction time of a subject to the decision about the third word.

The most important result reported [185] is that in the incongruent case with pattern masking of the second word, there is observed almost as much shortening of the reaction time to the third word as there is in the unbiased case. On the other hand when the second word is not masked then there is a delay in the response to the third word for the incongruent case in comparison to the unbiased case. In other words both interpretations of the second word palm (hand) and palm (tree) are accessed in the pattern masked case, and the relevant word is able to be used in the incongruent case, to speed up response to the third word. However, when consciousness supervenes then only the most appropriate of these meanings of the ambiguous second word is available for speeding further processing.

The simplest model of these interesting results, along the lines already hinted at [185], is that input to the semantic memory (SM) activates all possible interpretations of a given word. These may give rise to lateral 'spreading activation' to nodes representing similar meaning, but not to any spreading inhibition. This only occurs when access is being made to the working memory (WM) net. There lateral inhibition is used in the process of competition on the WM net, in order for a winning node to first reach a given criterial threshold. This competition may also involve more global competition using feedback from WMs in other codes or episodic memory. Once the winner emerges then the breakthrough to consciousness takes place. This may be broadcast to the other WMs and to the episodic memory (although the latter may not play a very important role in the short term, since consciousness is experienced by those who have lost their hippocampi).

It is possible to construct a neural model of this process of SM $\longrightarrow$ WM, with competition on the WM, to attempt to derive the more detailed, and especially the quantitative, features of the results. This has already been initiated [311], where the equations of the SM and WM where written down for the various processing situations. They were then analysed in terms of the control parts of the experiment [185], and a perturbation method developed about them. This leads to a deduction of relationships between various of the reaction times of the experiments, and a reduction of the total of 18 separate reaction times reported to a set of five underlying parameters. These latter are the speed-up or delay times which lateral connections (from SM to WM) bring about. In particular it is essential that there be lateral inhibition on WM, although this does not seem at all necessary on SM. In our current work an initial simulation has been done which bears out the above analytic results [122]. The basic model is thus

$$\text{IN} \longrightarrow \text{SM} \longrightarrow \text{WM} \longrightarrow \text{winner}.$$

One of the basic puzzles of such a model is the apparent duplication of the information in SM and WM. The only difference between these two modules is the presence of lateral excitation on the WM, not present on SM. Why should there be such duplication, apparently in a number of modules in the cortex? This is a question already raised by psychologists involved with working memory. An answer may be suggested from the hint of the layered structure of the cortex: SM and WM are the layers 4 and 5, and 2 and 3 respectively. This identification stems from the level of inhibition present in the upper layers and the different processing times between the upper and lower layers observed in various experiments [190]. It is clear that such a proposal sheds a different light on the nature of the SM and WM modules, leading to the question as to how they became differentiated from the surrounding cortical regions (if that also is the most suitable picture of the system). In any case, the suggestion does allow for a much more detailed exploration of the functionality of cortical layers than has been available heretofore.

7.7   Return to consciousness

The manner in which the feedback memories, aroused by input, are to be used must now be reassessed. The first way that this occurs will be through the encoding of the input at a semantic level. It was noted earlier that the nature of semantics is something of a mystery, but at least there are simplified neural network models of semantic networks, in which dedicated nodes exist to represent the meaning of each word. Such a network can be used to perform the initial encoding of the input. The updating of the semantic networks is also problematic, but very likely occurs during sleep [312].

A further influence of the records of the past is for direct feedback of autobiographic memories to the input area where the inputs are still active, as was suggested in section 7.4 and is observed experimentally in the late visual area. This could be of value in any verification [146] or decision mode of operation [307], where direct subtraction could be performed. This could occur, for example, in layers 2 and 3 of cortex from feedback activity from higher areas, possibly with support from the thalamic–NRT–cortical system of section 7.3.

If feedback is used to 'flesh out' the input in some manner, there is danger, however, of hallucinations or distortions of the input being experienced if this amplification becomes too strong. This may be avoided by temporal lobe memory excitations (both episodic and semantic) also being used in parallel to the input for later processing. Indeed, such a possibility leads to a clear distinction between the checking mode of operation mentioned above and what may be termed the constrained parallel mode. In the latter, parallel memorial activations lead to extra constraints on future actions and internal limbic or other responses, but not necessarily to modified inputs *per se*. They would therefore be still available for verification.

Such a difference in use of feedback is necessary in order to be able to handle the information being supplied by the memory activations. Even if predictions or verifications are going according to expectations, there is still a need to make further predictions and to provide additional information, along with the initial input, so that it can be efficiently manipulated to achieve whatever goals of the system are presently most important.

7.8 Personality and self

One of the most crucial aspects of self could be arguably claimed to be the emotional content given to experiences by the neural modules comprising the self. One of the most important of these is known to be the amygdala, an almond shaped set of nuclei in the mid-brain and next to the main organ of episodic memory, the hippocampus. It is possible to model the value which is attached to an input by a suitable memory representation in the amygdala. Emotions are very likely aroused by feedback from the amygdala to almost the whole of the cortex, especially to the sites of working memory, so colouring ongoing conscious activity. In this manner the 'concern' aspect of emotions can begin to be incorporated.

However, there is more to the self than simple colours added to consciousness. There is self-memory, to allow the persistence of self, as well as the ability to introspect in an active manner. This ability to process actively also has to be analysed.

The frontal lobes of the brain are now accepted as the sites of active processing, such as decision making, thinking and reasoning, controlling and monitoring the use of automatic 'schema', and attention. There is increasing understanding of the neuronal architecture of the frontal lobes which might allow such an important range of activities to be carried out. In particular there are at least five great loops of neural connections which parcellate the frontal part of the brain into different regions. One of these is involved in motor control, another in developing social interactions including emotional colour, a further one in reasoning at a low level, a fourth at a higher level, and finally one concerned with the control of eye movements.

It is that loop of neural connections which is involved with social interactions which seems to be also involved with self. There are parts of the loop in which there is laid down the autobiographical memories on which the self depends; the amygdala is also part of the system, and has already been noted as being involved with attaching value to inputs.

A number of neural models have been proposed to explain the activities of this and related loops in the frontal part of the brain. One is the ACTION network, which allows a simple explanation of the manner in which:

- Continued activity of neurons occurs over extended times (up to thirty seconds or so).
- Comparison between activity held on the frontal lobe with new input is achieved.

The connections of the ACTION network are modelled on those already known to be present in the frontal lobe; as such it presents an interesting possibility for explaining high-level cognitive function.

Using the above features of the ACTION network it is possible to begin to understand some of the features of self. Thus it is reasonable to consider that one of the important functions of self is of monitoring ongoing activity, to check that it is achieving the desired results. That mode of action could be achieved by the use of the ACTION network as a continual monitor of responses and inputs, in relation to predictions as to what they are meant to be according to autobiographic memories aroused by the inputs and past activities in the frontal lobes. This monitoring process can be applied to the activity of monitoring itself, but causes the whole system to crash since there will be no input to the monitoring unit when there is only monitoring of it by itself. That disappearance of self on attempting to regard oneself closer is one noted by philosophers in the past.

7.9   What is it like to be?

It is appropriate to turn to this question [203]: How can we ever know what it is like to be a bat? As Nagel argues so persuasively:

> ... every subjective phenomenon is essentially connected with a single point of view, and it seems inevitable that an objective, physical theory will abandon that point of view. ([203], pp 171–89)

As has been discussed in a number of contributions [81], there are a number of points being made here, which it is clearly impossible to consider at length in this chapter. However, the main result we have arrived at, adding to the earlier work of the author [309], is that a 'point of view' of any conscious being is determined both by its semantic and its episodic memory stores in a relational manner, as part of the relational theory of the mind [306].

In detail there are two sets of relations or constraints which the memory structure contributes to any input. One is that arising from the semantic memory related to the WM of a given modality. Such memory gives an automatic, pre-aware form of relational structure to input coding, which results in a general species or culture-specific point of view. Thus the input of a stream of sound leads very likely to activation of phonemic nodes which then activate nodes sensitive to certain words. Each word or phrase will itself have a semantic structure imposed on it by means of its relations to other word or phrase centres activated as part of the meaning of the word or phrase. Thus the word 'dog' would activate other word centres, such as those for 'walk', 'lead', or the type and name of one's own dog. These latter centres of activity would not necessarily be so active as to become conscious (although that could happen), but in general would give constraints to further brain activity in terms of the further words made more sensitive by this relational activity (in a predictive manner).

A second, more personal form of coding is that arising from the episodic memory of each individual. That would correspond more closely to the insider point of view, and would give more of the phenomenal content of conscious experience than the semantic memory. It would be the episodic memory of personal experiences with objects or other people which would give the more complete 'colour' to the inner life that would be so much more difficult to probe than the more objective semantic coding.

Could one build a virtual reality machine to give one the feeling as to what it would be like to be an X? One could provide an X's retinal transform on inputs, then an X's cortical transform, etc. Continuing in that vein one can see that the virtual reality machine would have in the end to perform a trepanning operation and replace the wearer's brain by an X's brain. In conclusion the best that one can do is build a machine, based on the relational principles above, with a structure as close as possible to that of an X's brain, and see if the machine responds as one would expect an X to. It is not possible to be an X, to experience 'from the inside' what it is like to be an X, but it is possible to understand an X's point of view, from the outside, by recreating that 'inside' relationally. If we know all of the relations operating in an X's brain then we know (scientifically) all there is to know about that inside, and certainly more than the X will ever know consciously by internal verbal report [312].

7.10   The PSYCHE project

We have presented two 'laws' of sensory awareness, in section 7.3, as deduced from the experimental data of Libet and his colleagues. These laws were then shown to be deducible from the NRT-complex competitive net model of conscious control [307]. More details of this latter were then filled out, especially so as to enable more detailed temporal features, associated with 'backward referral in time' of the experience, to be explored.

A particularly important question is as to the possible sites of injection of the cortical current which will gain control of consciousness. One proposal is that this occurs either at primary cortices (by injected current) or at sites of working memory, as mentioned in the previous sections. These latter would thereby allow activity to persist long enough to reach awareness, and also are known to be crucial in long-term memory. Thus an experimental prediction from such a hypothesis would be that suitable injected current in working memory sites might either control awareness or could disrupt it in the appropriate modality, as in the unattended speech effect [20]. Another experimental prediction is that there should be a gradual increase of neural activity in the region of NRT relevant to the appropriate winning site of working memory. As applied current is fed into cortex by the surface electrode this activity may be distributed in various cortical areas, but there must be a discernible increase of such activity at some sites in order for the hypotheses presented above to be valid. One place especially suggested in this chapter, as already noted, is the NRT, but it may occur elsewhere. There are many further predictions which may be made from this model [307].

The most important way that the above model can be fleshed out and at the same time a more effective understanding of the enormous wealth of experimental brain activity begun to be co-ordinated is by attempting to build a global model of the brain. That is one of the proposals of the PSYCHE project which is yet in its early stages. The programme is to construct the order of a hundred modules of simple neurons and connect them together in a fixed manner. There will then be an attempt to so arrange the connections and positions of the module as to obtain reasonable agreement with the results flooding in from non-invasive instruments (EEG, MEG, PET and MRI) measuring global brain activity.

7.11 Conclusions

A brief outline has been given of the relational mind model as a realistic neural network model of consciousness. It has scientific status in that it makes many predictions and opens up many avenues for further exploration. Whatever its fate there will be numerous related models suggested and tested over the next few decades as the programme of brain research becomes ever more mature. This leads the way ultimately to intelligent machines with our level of ability, and even beyond. The 'intelligent agent' of the future is now beginning to make the first few steps to start moving out of the regime of the science fiction writer and into the domain of the scientist.

Chapter 8

Neuroconsciousness: a Fundamental Postulate

Professor Igor Aleksander
Imperial College, UK
i.aleksander@ic.ac.uk

8.1 Introduction

There is a legitimate attack among those who write about consciousness [252] on the idea that there may be 'computational' models in the offing, as for example, advocated by Dennett [83]. The problem with these attacks is that they are usually made on rule-based attempts at modelling and these focus on the untenable position of a programmer who first has to work out what consciousness is and then write algorithms which model it. The main objective of this paper is to argue that neural systems provide an alternative which removes the programmer in all but the most trivial sense of having provided means whereby one can simulate things. This chapter discusses a fundamental postulate which relegates an artificial kind of consciousness to the working of a neural state machine. This is part of a system which consists of not only the fundamental postulate, but also 12 corollaries which have been listed elsewhere [148, 149]. It provides a justification for the fundamental belief that in neural state machines we have a computational medium which does not fall prey to the attacks mentioned above and which may be the right and proper medium within which to cast a computational theory of consciousness.

8.2 The fundamental postulate

The personal sensations which lead to the consciousness of an organism are due to the firing patterns of some neurons, such neurons being part of a larger number which form the state variables of a neural

state machine, the firing patterns having been learned through a transfer
of activity between sensory input neurons and the state neurons.

A postulate is a belief. In the context of consciousness, it is a belief which
is meant to be the answer to the question, 'What generates consciousness?'.
Clearly, as it stands, the postulate sounds a bit technical, and it is one of the
objectives of this chapter to clarify the technicalities. But the key implication
of the postulate is that consciousness is generated by some kind of machine. In
a living organism this machine is called the brain. But prejudice and tradition
has it that other machines being 'inanimate' objects (things without a soul or
'anima') makes it 'wrong' to use words such as 'consciousness' and 'machine'
in the same sentence. So, right from the start, I need to justify the notion that
consciousness is not outside the scope of what can be gained by studying objects
manufactured by man rather than God or Nature. I do this not by arguing that
man-created artefacts are necessarily as conscious as humans, but that by looking
at brains with the tools that are available for looking at machines, it becomes
possible to glean an inkling of what consciousness might be.

The idea that consciousness could be explained in this way is what
philosophers would call 'reductionist'. This is a polite way of saying that
the discussion has been simplified to such an extent that the explanation may
not refer to consciousness at all. But the almost hidden implication of the
postulate that consciousness may be simpler than many are prepared to believe
is raised in the belief that simplification and clarification go well together if
adequate precautions are taken. Non-reductionist philosophy often does not
satisfy those who have a curiosity about their own sense of consciousness. Fear
of simplification has deep roots in the belief that consciousness is something that
inspires great awe and must therefore be complex and, in some philosophies,
mystical. The fundamental postulate is purposefully designed to encourage a
view of consciousness which stems from being confident of knowing what being
conscious feels like and suspending any feelings of contempt one may have for
machines or simple, but relevant, logical arguments. This may be reductionist
in the sense that it simplifies things, but, I shall argue passionately, it does not
remove any of the awe one might have for one's own consciousness; it is meant
to replace mystique by logic.

Two questions are raised and then answered in this chapter. First, how
much of our individual curiosity about consciousness could be satisfied through
the study of the behaviour of an artificial, neural device? Second, how far
have the existing explanations of philosophers, biologists, linguists and computer
scientists gone towards this satisfaction? The happy conclusion is that nothing
in the fundamental postulate turns out to be too startling.

8.3 Defining my own consciousness

The only consciousness that I can be absolutely sure of is my own. So any
attempt at defining consciousness will start by having an intimate character,

the character of a confession or a personal diary. It is to stress this point that philosopher Thomas Nagel [304] posed the celebrated question 'What is it like to be a bat?'. Nagel argued that no amount of knowledge of the neurophysiology of a bat will lead a human to know what it is like to be a bat without being a bat. He concluded that neurophysiology is inadequate in giving a full account of consciousness. I believe this account to be wrong: the more we can relate our own feelings of consciousness to our brain mechanisms, the better we may be able to appreciate what it is like to be a bat. Fascination with consciousness, despite its very personal nature, is shared by a large number of humans and has been so over the ages. Since the beginning of recorded philosophy, thinkers have tried to share their views of what consciousness is with a wide audience. Inevitably these discussions have started with each philosopher's own thoughts about being conscious. Therefore it seems quite proper for me, even if I wish to rely on logical argument, to start with this introspective view of my own consciousness. I then guess that a reader will be able to relate what I say to their own feelings. I may not be able to know what it feels like to be a bat, but I may be able to speculate on what a bat may be able to feel, once I can work out how what I feel is based on what I know about my own brain. The first step is to try to articulate what precisely it is about my own consciousness which creates this curiosity.

Psychologists talk of the conscious state in contrast with the unconscious state. Being conscious means being awake as opposed to being anaesthetized or just fast asleep. When I am awake, being 'me' comes from the major sensations of my knowing who I am, perceiving and understanding where I am, knowing what I have done in the past and what, according to my current perception I can do in the future. I also can discuss all of this, and much more with those who share my language. The property which is common to all these attributes is 'knowing', a knowing which stretches from the very personal which I may choose to share with nobody to the knowing of language or knowing how to bake a good apple pie. What I would like to know, what needs to be explained about consciousness, is how this personal 'me-ness' relates not only to my brain, but to the rest of my physical makeup and the nature of the world I live in.

The intriguing 'me-ness' that makes me curious about consciousness, is what, in the first phrase of the fundamental postulate, I have called 'personal sensations'.

8.4   The consciousness of others

The hunt for consciousness so far seems to be a highly introverted affair. This is intended to be just a start which needs development and amplification. In discussions about consciousness it seems important to include any object which could possibly be deemed to be conscious. So, in the fundamental postulate, the word *organism* is used to include objects other than ourselves. Is an ant conscious? A cat? A pencil? The universe? Using the word *personal* is not intended to refer only of 'persons'. It points to ownership and individuality, in the sense that an ant could have a personal way of reacting to an obstacle. By using *personal* in conjunction with *sensation* and *organism*, I include objects that primarily must be capable of sensing things not only externally but in some internal way too (much will be made of internal sensing later). Clearly an ant has sensations which it owns and are therefore personal to that ant, but the same may not be true of a pencil. Why would we shrink from saying that a pencil senses the paper on which it is made to write? The difference between an ant and a pencil is that the antenna of an ant is a sensor which translates pressure into an electrical signal which can lead to a change in what the rest of the ant is doing. It may change direction or speed. Importantly, the ant is capable of an outward action and the outward action is dictated by some condition of the ant's internal machinery. So sensing leads to a change of state of the inner machinery which may lead to a change in outward action. Figure 8.1 illustrates these components.

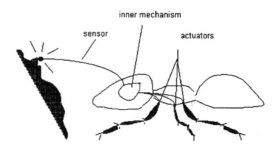

Figure 8.1. Sensor (antenna), mechanism (brain), actuator (legs) arrangement for an ant.

I suggest that in order to be scrutinized for some kind of consciousness, organisms must, minimally, have some inner machinery which may be controlled by sensors and which is capable of acting in some physical outward way. The outward action in living organisms is usually due to the contraction of some muscles which work in the opposite way from sensors. Muscles receive electrical signals and turn them into physical actions. Engineers call such outputs 'actuators'.

This broad definition lets in clockwork mice, ants, cats and Uncle Harry. A point that I shall make very often is that any study of consciousness should allow a broad range of organisms, but, at the same time, enable us to say with clarity why the consciousness of a clockwork mouse is pretty trivial while that of Uncle Harry very impressive.

8.5   The power of modelling

What is beginning to emerge from the above is the idea of a 'model'. Sometimes in science, but principally in engineering, models are used to test an understanding of a phenomenon by means of a kind of mock-up. For example, were a new bridge to be built, the civil engineer may think it wise to build a scale model to test whether the bridge will stand the predicted loading. Clearly the model bridge does not serve the same function as the real thing, its main function is to test whether the way that bridges work has been properly understood. A model need not be a scaled version of the real thing, it could, for example be a simulation on a computer. This would consist of a major program which contained smaller programs for the behaviour of all the components of the bridge and the effect that they have on one another. When the major program is made to 'run' it would coordinate the smaller programs and invite the user of the machine to specify a load for the bridge. The program would calculate the effect of the load on all the components and show the amount by which the bridge would deform under the load and whether there might be a collapse. Indeed the bridge could be tested to destruction without incurring the cost of any damage.

The model would, by some, be seen as a list of mathematical laws that govern the components and their interactions. Mathematicians would call this a set of equations, with which, given the value of the load, they may be able to work out whether the bridge will stand up or not.

When it is said that a minimal list of things that is required by an organism which is to be scrutinized for consciousness is an input sensor, an inner mechanism and actuators, this spells out a framework for a model which will apply to all such organisms. The distilled nature of this framework is shown in figure 8.2.

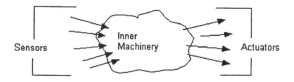

Figure 8.2. Framework for a model.

The vague part of this model, and the part which will enable us to talk of the sophistication of the modelled system is the 'inner machinery'. A neat trick that engineers have for modelling even the most complex of systems is to think of the inner machinery as being capable of being in a number of 'states'. Part of the inner machinery of a car, for example, is the gears. A five-gear car has five gear states: each such state determines a different way in which actuators (the force on the driving wheels) respond to sensors (force on the accelerator pedal, the weight of the car, the slope of the road, the angle of the wheels and many other factors).

The beauty of this approach is that any complex mechanism can be modelled by this technique. For the moment, what has been said only explains the appearance in the fundamental postulate of the words *state machine*. The model framework of figure 8.2, when the inner mechanism is represented by states, is a state machine. The words indicate that the general power of such state machines will be used as model which, as in the case of the model of the bridge, will give us a way of checking that what is being said about brains, minds and consciousness is consistent and coherent.

8.6   The danger of inappropriate modelling

There are appropriate and there are inappropriate models and all of them could be run on a computer. I do not mean that a model could be just wrong. What I do mean is that a model can arrive at the right answer without providing the understanding it was meant to provide. This is the sad story of a field called symbolic or classical artificial intelligence (AI). Defined as 'doing on computers that which if done by humans would require intelligence', AI occupied researchers in computing laboratories in the 1970s. AI programs were meant to be models of human cognition, that is, human thinking. Probably the most familiar product of this work is the chess playing machine which every respectable 'executive toy' shop will sell. These systems are based on a programmer's notion of how a good game of chess might be played. The computer is made to look ahead of a particular state of the chessboard and work out every possible move for several steps of the game. The consequent board positions are evaluated against a list of benefits and the most advantageous available move is taken. While such machines can easily beat me, they must look at a large number of steps ahead and evaluate a prodigious number of board positions to beat a good player.

While such machines have their fascination and merit, their designers cannot pretend to have made models of human thinking. On television, champion players are sometimes asked to comment on the replays of their games. They say things such as '... he's got me in a tight corner now... but I think I can win if I keep on fighting for seven or eight moves... I could sacrifice a pawn to give him a false sense of confidence... I know he doesn't like using his queen's bishop... '. This is no mechanical evaluation of moves. The player is using a wealth of past experience, cunning and even a psychological knowledge of his opponent. So, while the AI program plays a game which humans play, it is not possible to say that it is a model of a human player. It is merely a programmer's idea of how a machine might play the game. It is for this reason that AI sceptics will say that a computer model which absorbs and stores a story represented in natural language, and then answers questions, does not necessarily understand the story. This is the nature of Searle's objection to AI. He argues that a programmer can write complex programs which are very good at manipulating words by means of many rules, without capturing what is meant by 'understanding'. Given the Humpty Dumpty rhyme, if asked 'why could he not be mended', I would probably answer 'because he broke into too many pieces'. I do this by appealing to my experience and knowledge that Humpty

Dumpty was an egg shell and egg shells shatter when you drop them. The computer would have to have rules such as 'fragile things are hard to mend' and 'Humpty Dumpty was a fragile egg'.

Some commentators [252] take this argument to mean that any model that runs on a computer cannot be an appropriate model of consciousness, as running on a computer is in a different category from living and knowing. I argue that this is not so. The examples above are inappropriate, incomplete models because they do not have their own way of building up and accessing experience. This does not mean that appropriate models do not exist. Appropriate models of the mechanisms of consciousness have to answer the question of how consciousness is built up and cannot simply be AI mock-ups with mechanisms supplied by a programmer's idea of what consciousness is. It is this need for appropriateness which leads to an appeal to neural models in the fundamental postulate as is explained below.

8.7   Why neurons?

We come back here to the idea of inner states. What are they in a living brain? The building brick of the brain is the *neuron*, a cell which receives signals from other neurons (input) and transmits a signal to further neurons (output). This is illustrated in figure 8.3.

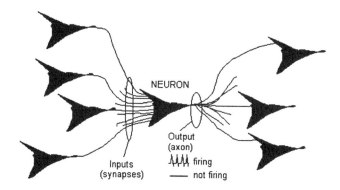

Figure 8.3. A neuron in a net of neurons.

Without wishing to minimize the beautiful and detailed electrochemical properties of neurons found by neurobiologists, the essence of what a neuron does can be described very simply. At times, the neuron generates at its output (axon) a stream of electrical impulses (about 100 impulses each second) and, at other times, almost no impulses at all. It is said that the neuron either 'fires' or 'does not fire' to describe these two conditions. Whether a neuron fires or not is determined by the current pattern of firing and non-firing of the 'input' neurons (where the input connections are called *synapses*). Which input pattern leads

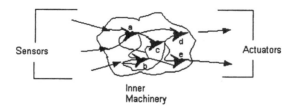

Figure 8.4. A neural net model.

to firing and which does not is something that the neuron can learn (explained below). Figure 8.4 shows the way in which a network of neurons could be seen to form the interior of the model in figure 8.2.

Labelling the five neurons as 'a', 'b', 'c', 'd' and 'e' it is clear that the actuators can only react to what 'd' and 'e' are doing. The others, 'a', 'b' and 'c' are the inner neurons which create the internal states. How many of these states are there?

As each of the inner neurons can either fire (f) or not fire (n), 'a', 'b' and 'c' can be in eight states: fff, ffn, fnf, fnn, nff, nfn, nnf, nnn. The number of inner states can be calculated for any number of neurons N (it turns out to be 2^N). These inner states are particularly interesting because of the inner network of connections. This makes 'a', 'b' and 'c' interdependent, it gives them a behaviour 'of their own'. This property is thought to be essential to be able to represent concepts such as 'thinking'. Even if the sensory inputs of the organism are unchanging and the effect of the 'thought' is not immediately discernible at the actuators, the inner states can change. That is, the model can represent private, inner events. Each of the inner neurons has been called a *state variable* in the fundamental postulate which is meant to suggests that much of the privacy that goes with consciousness, much of the inner thought which is so central to the mental makeup of a living organism, can be modelled by the activity of inner neurons.

Returning for a moment to the number of states of a neural mechanism, it is salutary to think that the human brain has ten billion neurons which implies that the number of inner thoughts a human being can have is vast. Imagine for the sake of argument that a human has one hundred 'thoughts' (inner states) a second. In a lifetime of, say, 100 years this would amount to a total of 316 billion thoughts. The amazing implication of the 2^N rule is that only 39 neurons are required to have that number of states! What's wrong with this calculation? The problem lies with the coding of these states. The 316 billion states are all the possible patterns that could appear on a row of 39 lights, each of which is either on or off. All that this calculation does is to show that a few neurons can have many states, but the states on these 39 lights are not terribly interesting or meaningful.

The fact that there is a large number of neurons in the brain points to

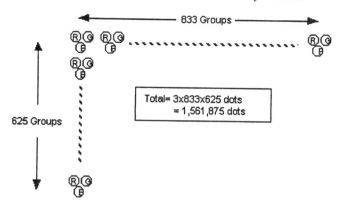

Figure 8.5. Dots on a television screen.

the idea that each thought could be encoded in a much richer, more interesting way. That is, with billions of lights one could represent colourful and detailed pictures as is done in the advertisements which, at night, adorn Piccadilly Circus or Times Square. Could each thought resemble a television picture? Let's pursue this hypothetical idea. The screen of a colour television set is made up of tiny groups of dots as shown in figure 8.5. Each group has three dots, one for each primary colour (red, blue and green).

Being generous, say that ten neurons are used to control the intensity of one dot: something of the order of 16 million neurons could represent a frame of a full-colour television picture. Now, the cortex in the brain (the outer sheet which is involved in sensory processing, particularly vision) has tens of billions of neurons with a density of about $100\,000$ mm^{-2}, the whole cortex spanning about half a square metre (the size of a large handkerchief) if stretched out. So, an area of the size of a thumbnail (160 mm^2) would contain the 16 million neurons capable of giving us an 'inner' TV show for a lifetime.

I hasten to add that I am not suggesting that the mental images which are so much a part of our conscious life are generated in a small area of the cortex. As Crick [96] points out, we may not know where neurons associated with conscious thinking can be found in a living brain. However, the point of all this is that it draws attention to the massive power of the neural networks we carry in our head which is such as to be able to represent a great deal of information including mental images of considerable richness. It is for this reason that in the fundamental postulate the *neuron* is referred to as a *variable* of a *state machine*. This suggests that the inner mechanisms of the brain have enormous potential for representation of past experience, understanding of current perceptions and predictions of the future. However, three major questions remain unanswered. How does an organism which owns such a representational brain also 'own' and internally 'see' these representations? How do the representations get into the brain? What does playing with artificial brains tell us about all this? The next three sections show how the fundamental postulate reflects these issues.

8.8 The mind's eye

If I sit in a darkened and quiet room I can think of many previously experienced events. I can remember things as different as last night's dinner, driving to work, Pythagoras' theorem or Mahler's second symphony. To me, my brain appears to be a veritable multi-media computer. Seeing and hearing appear not to go away just because my sensory organs are currently inactive. The fundamental postulate takes an extremely simple view of how this might happen.

If we touch something with a finger the pressure-sensing neurons in our fingers cause the firing of some neurons in our brains. It is this inner firing to which we give the name 'sensation'. It is well known from the anatomy of the brain that different neurons are associated with different sensory parts of our body, so if I touch something with my tongue, a bunch of neurons in my brain fires but this is a different bunch from that which fires when I touch something with my finger. In fact neurophysiologists have mapped out how the different touch-sensing organs of the body map onto an area of the cortex in the brain called the 'sensory strip' (figure 8.6).

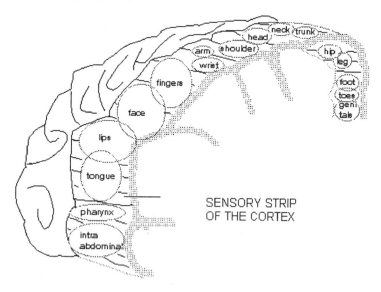

Figure 8.6. The sensory strip in the cortex.

Figure 8.6 shows this mapping in the left part of the cortex which corresponds to the right side of the body. There is a similar strip on the right-hand side of the cortex which corresponds to the left of the body. While it is not my intention to launch into a lesson on neuroanatomy, I merely draw attention to the very important property of the brain to reflect, in miniature, sensory patterns. However, when I touch something, while neurons are firing in my sensory strip, the sensation I have appears to be 'in my finger'. The reason for this is that while the touch sensors are active, much other firing in other parts of the brain is associated with that event. Many muscles are holding appropriate positions to get the tip of my finger to where it is. Every time that a muscle does something, it sends signals back to the brain which cause further firing of neurons in parts other than the sensory strip. We shall see later that these multi-channel patterns give us a sense of self: patterns in the brain are related to where bits of our body are and what they are sensing.

Now, say that one such pattern of inputs can be stimulated artificially, we could be fooled into thinking that we are feeling something very specific with the tip of a finger. Returning to the question of where pictures in our heads come from, it is merely necessary to think of the sensory retina at the back of the eyeball as a set of densely packed sensitive 'fingers' which cause patterns of neurons to fire in the visual part of the cortex causing a sensation we call 'seeing'. The same goes for smell and hearing and again some internal stimulation of these neurons could produce the sensation of seeing or hearing in our heads. There are many stories told of the reports of patients whose brains are artificially stimulated during brain surgery, they hear symphonies, see visions and smell flowers. In other words, in common with all living organisms we *own* a representation of the world as specific combinations of the firing of neurons in our brains. The key property of such patterns is that they are self-sustaining for a while, and may be triggered by sensory experience *or* other inner events. That is, we could report a particular firing experience as 'seeing' a dog or 'thinking' of a dog depending on whether the trigger was external or internal. It is not being said that the acuity of the two patterns is the same, we see a dog with greater clarity and specificity than we think of a dog. On the other hand we can think of one of several dogs we know. We could even imagine a dog we have never seen before. Much more can be said of the way in which such rich and varied representations can coexist in one limited piece of neural machinery. Here I merely stress the importance of the use of the phrase *firing patterns* in the fundamental postulate. Next, I need to say something about how such firing patterns get into brains, a process which is sometimes called 'learning'.

8.9   Learning and remembering

A primary property of a neuron is that, in some raw, 'inexperienced' state it can be made to fire by incoming stimulation. For example the firing of a sensory nerve in the tip of the finger caused by pressure makes a neuron in the sensory strip (figure 8.6) fire, a bit like a bell ringing inside the house when the doorbell is pressed. Learning consists of the neuron 'taking note' of the pattern present at its other inputs. Learning is complete when the firing of the neuron is controlled

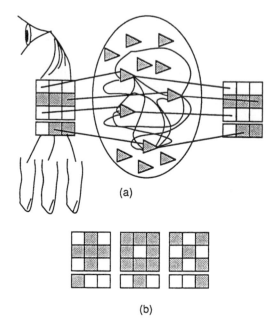

Figure 8.7. (a) The net and its connections to fingers and an eye, (b) three learned sensory experiences.

by the patterns the neuron has 'taken note' of. With figure 8.7 I shall try to explain how several neurons may be working together to 'remember' patterns of sensory input. The figure represents a system with three 'fingers' and an 'eye'. This is a purely fictitious system, nothing *exactly* like it is likely to exist in the brain. But things *a bit* like this do go on in the brain, and according to the fundamental postulate, underpin conscious experience.

In figure 8.7(a), the arrangement of squares on the left is merely a way of showing what various sensory elements are doing. The bottom line shows shaded squares where the sensors in the fingers are causing firing to enter the net. That which is sensed by some kind of 'retina' in the eye is shown as a three by three arrangement, again shaded squares representing the connections which are firing (a horizontal bar is being seen). All these sensory connections enter the neural net (for which only some of the connections are shown). Any neuron in this net which receives a firing signal is assumed itself to fire and the firing of the net neurons is shown in the arrangement of squares on the right. It comes as no surprise that the pattern on the right is the same as the pattern on the left. The net seems to be just a relay station which passes on information it receives. But it is here that the interconnections *between* the neurons come to play an enormously important role.

It has been said that any neuron which fires 'takes note' of its other inputs.

Because the other inputs come from neurons which are triggered by other parts of the input pattern they actually represent part or all of the input pattern (depending on the number of inputs to each neuron). So each neuron, through the fact that it will fire when *almost* all its input conditions are the same to those that have been learned, will fire in consensus with other neurons for learned patterns. All this means that any pattern (such as the three examples in figure 8.7(b)) which is sensed while the neurons are learning will become self-sustaining in the net. That is, the learned patterns can appear in the right-hand set of squares even if they are not sensed. What is more they can be made to appear by partial firing patterns at the input. In human terms, this is like saying that the cross seen by the eye in figure 8.7(b) is 'felt' by the left finger, the square by the middle finger and the X by the right finger. The fact that the net can sustain these combinations only, means that if the right finger is stimulated and the other two not, the net will reconstruct the associated X within itself and 'see' X but only with some 'inner eye'.

Not all the ins and outs of this fascinating property have been explained here. The mechanism just described is merely meant to explain why the following words have been used in the fundamental postulate:

The firing patterns having been learned through a transfer of activity between sensory input neurons and the state neurons.

8.10 In a nutshell...

It is now worth revisiting the postulate in a way which is less terse than that found at the beginning of this chapter. Consciousness, according to the fundamental postulate, has all to do with the sensations which each of us has within our heads. The key submission is that such sensations are due entirely to the firing of some neurons. But the firing of these neurons is highly conditioned by sensory experience. A neuron is not only driven by the perception while the act of perception is in course, but also takes note of what many other neurons are doing at that time. As these other neurons also represent other parts of the perception, learning creates self-sustained patterns in networks of neurons and these patterns have an 'internal' existence of their own. They can be triggered by part of the perception such as 'touch' leading to the internal 'vision' of what is being touched. A distant roll of thunder may make us conscious of impending danger or, being more poetic, the smell of a flower in spring may make us conscious of the longing for someone we love. Association is a highly important mechanism which comes naturally to networks of learning neurons. It has also been said that the sensations which neurons can produce can be astronomical in number and highly complex in their interrelations. Some may call the fundamental postulate 'oversimplified'. I recommend to the reader to embrace simplicity as a friend: it does not detract from the sense of awe and wonder at the way in with which a neural net can conjure up both the beauty and the usefulness of a rich mental life. So far, only the surface of an explanation has been scratched, more questions have been raised than answered. What is awareness of self? What is free will?

There is, however, one final question of some importance which needs to be answered in this chapter. If consciousness is really generated by the simple mechanisms suggested by the fundamental postulate, why has this not been obvious to thinkers who, both over the centuries and in recent times, have puzzled over what the phenomenon is or is not? The answer may be that the idea of a 'state machine' is relatively new (40 or so years old). This is not very long in a history which stretches back for several millennia. Also, while the principles of state machines are commonplace for some computer scientists and engineers, they do not feature in the thinking of many modern philosophers and others who have attempted to describe consciousness. They are treated with suspicion and put in the class of modelling through the use of 'golems'. Indeed, some contemporary philosophers [249] have an explicit distaste for this kind of modelling as, they argue, the model being synthesized is devoid of the very thing one is searching for: an independent existence of 'coming into being'. The model has a 'constructed' identity which makes it invalid. It seems important to keep an open mind on this issue and not hold back from gaining insight if such insight is in the offing.

Acknowledgment

Our thanks to World Scientific Publishing for giving permission for reproduction of some of the material in this chapter (from *The Impossible Mind of MAGNUS* [149]).

PART 3

ADAPTIVE ROBOTICS

We are all familiar with images from industry of robots assembling cars, welding and doing other seemingly complex tasks. However, these robots lack the power of adaptation. If their environment changes, such as by the misplacement of some product or piece of equipment they depend on, they fail ungracefully. This lack of adaptation restricts them to rigid, pre-set environments. In contrast, robots using neural computing techniques are able to adapt to their environment, with knowledge of the environment not necessarily being built into the robot when it is constructed. This part of the book describes several approaches to making robots more flexible in their behaviour by building into them the power of adaptation. There is also a discussion of how adaptive robots can be built using biologically realistic training algorithms.

Chapter 9

The Neural Mind and the Robot

Noel E Sharkey[1] and Jan N H Heemskerk
University of Sheffield, UK
[1]n.sharkey@dcs.shef.ac.uk

9.1 Introduction

I got no strings, to hold me down, to make me fret, or make me frown.
I had strings but now I am free there are no strings on me. [From
Disney's *Pinocchio* (1938), *I've Got No Strings* (Washington/Harline)]

Since the time that 'God made man in His own image', humans have been
fascinated by stories about artifacts coming to life. Ancient myth, fairytales,
literature, and science fiction abound with stories of artificial beings. In the
older stories, although it is people who make the beings, it is the power of
supernatural forces that bestows life. Ovid's story of Pygmalion is perhaps one
the most famous from mythology; a sculptor falls in love with his sculpture of
a woman which the goddess Venus then brings to life. Then there is the ancient
Guianan Indian fairytale about a witch doctor who carved himself a daughter
out of a plum tree because he needed a son-in-law to look after him. Similarly,
there is the story of the wooden puppet Pinocchio who desires and eventually
obtains boyhood (a desire that parallels that of the android Commander Data in
Star Trek: the Next Generation).

In days of old, the breath of life into the inanimate was a mystery that
extolled the powers of magic and gods. But now it is science that brings the
fantasy. To begin with, the creation of life by science had supernatural overtones
and Gothic consequences as in Mary Shelly's Frankenstein. Even the word *robot*,
from the Czechoslovakian word *robota* meaning compulsory labor, comes from
literature. It was first used by a Czech playwright Capek in his 1920 play
Rossum's Universal Robot about the development of anthropomorphic machines
as servants for humanity. The science fiction writer Isaac Asimov was the first

169

to describe robot technology as *robotics* and began the fascination of science fiction writers, film makers, and audiences with intelligent autonomous human artifacts and his three laws of robotics (see Aleksander and Burnett for a good overview of the history [8]).

Today it is commonplace to find robots and androids littered casually throughout any science fiction story. Almost everyone with a television set is familiar with the most sophisticated imaginable robots. Television series such as *Star Trek*, and Hollywood films such as *Terminator 1* and *2*, expose us to very complex man-made creatures or artifacts. We are not even surprised when a machine is indistinguishable from natural men and when a machine is ripped to pieces we expect it to run a self-healing procedure, get back on its feet and walk off the scene. From Robbie to Hal to the Terminator, robots have been empowered with super intelligence and special powers. We even get hybrid systems combining machines with living beings as in *Robocop* (or the Borg in *Star Trek*). But what are the grounds for this incredible speculation from the science of robotics?

Robots in the real world of industry are a pale reflection of their screen counterparts. They can rivet and they can paint, often with superhuman speed and certainly without boredom, but they are usually confined to a fixed point, with a limited scope for largely repetitive actions. They have no awareness of their environment, or understanding of the tasks that they are 'mindlessly' undertaking. They would happily follow instructions to paint a car even if there were no car present. Unlike Pinocchio, these robots do have (virtual) strings. They are like automated puppets; puppets equipped with a computer that has been precisely instructed by the 'puppeteer'.

An important goal of modern 'scientific' robotics is to *cut the strings* and give the robot its autonomy—'Hi ho the merio, that's the only way to be'. There are many differing motivations and approaches to the development of autonomous robots. One of the most common of these is the behavior-based approach [48]. The basic idea is to provide the robot with a range of simple behaviors and, extending our analogy, allow the environment to pull the strings. That is, the environmental conditions determine which behavior is active at any time. Others believe that the best approach is to provide the robot with a 'simple mind' [179], say the mind of an insect, and provide it with a means to learn from its environment [71].

In this chapter we discuss the move from the metal to the mental through the use of neurocomputing techniques and their property of learning from examples. We begin, in section 9.2, by taking a very simple beast, $Beast_1$, and using it to illustrate some of the important issues in modern autonomous robotics such as:

(i)   The nature of autonomy.
(ii) How we describe a robot's behavior.
(iii) The fundamentals of an artificial neural network mind.
(iv) The need for an autonomous robot to adapt to its environment.

In section 9.3, we catalog and describe some of the current research on training techniques for mobile robots. Finally, in section 9.4, we speculate about how the field might move forward from its present position to develop robots with artificial neural minds.

9.2 The environmental puppeteer

What exactly is meant by an autonomous robot? There have been many attempts at definition in the literature. The *Oxford English Dictionary* defines autonomy as:

> Freedom of the will; the Kantian doctrine of the self-determination of the will, apart from the object willed.

We certainly cannot expect this from our robots, but the definition of an automaton from 1611 is closer to the modern use:

> A piece of mechanism having its motive power so concealed that it appears to move spontaneously; now usually applied to figures which simulate the actions of living beings, such as clock-work mice, etc.

According to Bourgine and Varela [44], the study of artificial beings arises from

> ... the need to understand the class of processes that endows living creatures with their characteristic *autonomy*. Autonomy in this context refers to their basic and fundamental capacity to *be*, to assert their existence and to bring forth a world that is significant and pertinent without being pre-digested in advance.

One of the problems with the definitions of autonomy and, indeed, in the study of living beings, is that it is too easy to ascribe complexity to a simple mechanism. With the appropriate environment, a simple mechanism may exhibit an apparently complex behavioral repertoire. This has been argued forcefully by Braitenberg [45], a neuroanatomist, who showed how what appeared to be sophisticated and complex behaviors could be reproduced using a vehicle with only two sensors wired to two motors. He could create behaviors associated with fear and aggression, such as running from shadows, or hiding in dark places and pouncing on prey; even going as far as love. The following new example of complexity emerging from simplicity shares Braitenberg's philosophy and should make the idea clear.

9.2.1 *Beast*$_1$

Imagine that you are looking through a window into a darkened room. Someone switches on a light and you observe a small shape move across the floor and vanish under a low chair. Your curiosity risen, you enter the room with a torch (flashlight) and switch off the overhead light. You get down on your knees and shine the torch from different positions under the chair where you saw the shape disappear. Suddenly you hear a whirring noise and are approached quickly by a little beast that stops beside you. Before you have a chance to take a proper look at it, someone switches on the main light and it disappears under the nearest chair again.

What do you make of this creature? Is it an autonomous being? Is it friendly or harmful? It seems to dislike light and yet when you shone the torch under the chair it approached you. Did it understand that a person was holding the torch and it approached looking for food? Can it distinguish between torchlight and overhead lighting? You must try an experiment.

You switch off the main light again and shine the torch under the chair until you hear the whirring sound, then you set the torch on the floor, still pointing under the chair. The creature immediately approaches the torch and this time crashes straight into it. Is it showing anger at this intrusion? Does it know that you are no longer holding the torch? To finish the experiment, you switch on the overhead light again and the beast 'scurries' under another chair.

A second experiment is now required. It seems clear that the alleged beast can differentiate between an overhead light and a torch. But, more intriguingly, it appears to be able to distinguish between a torch being held and a torch lying on the floor. It 'attacked' the latter for a start. Maybe it just behaves differently towards the two because one was on the floor and one was not. You reach for your anglepoise lamp:

First you shine the lamp under the chair to 'attract' the creature and then you leave it with the lamp off the floor. The beast approaches, moves through the beam and stops in the shade. Aha! It seems to be a strategy to get out of the glare of the light. Rather than run away, it moves through the light and into the shade beyond. But then why did it 'attack' the torch? Perhaps it was trying to extinguish the light.

As a final test you pick up the lamp and move across the room. Oddly the creature follows you. You switch off the lamp and move away. Then you set up the lamp with the bulb pressed down on the floor, switch it on again, and move out of the way. The beast quickly approaches the light bulb, as expected, but surprisingly it does not 'attack' the bulb. Rather, when it gets close, it simply turns sideways to the lamp and stops in the light. So, it can tell the difference between a torch and a lightbulb on the ground. A complex little creature indeed, with a rather odd relationship to light.

At last you take a closer look at the beast 'basking' in the light and notice that it is actually a small mechanical vehicle with wheels and what appear to be sensors at the front. Now you have a good idea about how it works. It is one of those AI devices equipped with an onboard computer, you reason, without even considering a positronic brain. You wonder how it was programmed. Noticing that the top lifts off, you look inside. To your surprise there is no computer, no

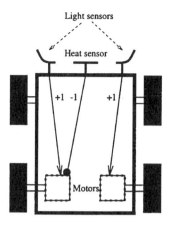

Figure 9.1. *Beast₁* and its wiring.

microchips, and no circuit boards. There is just a battery and *three wires*. What? You take the whole device apart, feverishly looking for a radio transmitter or some sort of remote control, but the search is in vain. It was the environmental puppeteer at work.

How can such a simple device produce such seemingly complex behavior? The vehicle, as shown in figure 9.1, has three sensors at the front, two light sensors and a heat sensor. These are all binary threshold sensors which means that when light (or heat) intensity reaches a certain value (threshold), the sensor outputs a +1. The values on the wires are weight coefficients. The value received by a motor is the value of the output from a sensor multiplied by the weight coefficient on the wire connecting motor and sensor. If this value is greater than a threshold of 0, the motor turns on. Thus when light shines on one sensor, one motor turns on and the vehicle turns toward the light; hence the whirring sound of the motor under the chair. When light shines on both sensors, both motors come on and the vehicle moves forward. With overhead lighting, this means that the vehicle will continue to move forward until it finds itself in the shade. With torchlight, it will move forward until it bumps into the torch, seemingly attacking it. There is insufficient heat to trigger the heat sensor. With the anglepoise lamp, the vehicle will move towards the bulb on the floor until the heat sensor is triggered or thresholded. Then one of the motors will be turned off (or inhibited) and the vehicle will turn to the side and stop—since its sensors are out of the light—seemingly 'basking' in the light.

9.2.2 Describing behavioral systems

Looking at the behavior of the *Beast₁*, many people would call it an autonomous robot. Seeing the three wires, however, it appears that its behavior is entirely determined by light and heat rather like the behavior of a plant.[1] In our puppet analogy, it appears to be the *environmental puppeteer* that is pulling the strings.

However, even if the innards of the beast and its environmental determinism are known, we may still attribute much to its behavior just because it moves without us seeing the strings. As Skinner [287] put it:

> Anything which moves is likely to be called alive—especially when the movement has direction or acts to alter the environment.

Skinner tells the story about two of the great romantic poets passing a steam engine in the 19th century:

> Wordsworth observed that it was scarcely possible to divest oneself of the impression that it had life and volition. Coleridge replied, 'Yes, it is a giant with one idea.'[2]

It is the observation of behavior (and the entertainment of certain philosophical beliefs) that determines whether or not we think that an organism or mechanism is autonomous or deterministic. We qualified this statement with 'philosophical beliefs' because many philosophers have held the belief that all animal behavior (or all except humans) is entirely determined by forces external to the animal. It really has to be said that autonomy and behavior, like beauty, are in the eye of the beholder. However, for a scientific appraisal such statements leave much to be desired.

The behavioral rules of *Beast*[1] can be defined in a look-up table as shown in table 9.1. This shows the precise input/output relations of *Beast*[1], although it may be difficult to work these out through observations in a complex environment. Moreover, as we have discussed, it would be extremely difficult, if not impossible, to work out how the control system was implemented. These are the sorts of behavior that we could determine experimentally by turning the vehicle on its back, presenting heat and light sources to the sensors, and then observing the movement of the motors. This description is similar to what Keijzer and Heemskerk [159] refer to as a *proximal* description of the behavior. This they contrast with distal descriptions, which we shall turn to momentarily.

When we place the beast in a real-world environment we say that it is *situated*. The proximal description of such a situated robot consists not just of sensor and motor states, but also of a movement. A single sensor state resulting in a motor state resulting in a movement is called a sensory–motor loop. The movement is important in the proximal description in that it may result in new sensory states (this does not happen with the robot on its back on the table),

---

[1] The first mechanistic simulations of this type were conducted by Loeb [180] with his work on tropisms and a mechanistic approach to zoology. See also [330] for early work with an electromechanical tortoise.

[2] This is reminiscent of the strong AI position of John McCarthy of Stanford University who says that the thermostat in his house has three 'beliefs': it's too hot in here, it's too cold in here, and the temperature is just right in here.

Table 9.1. A behavioral look-up table for *Beast₁*. The subscript *s* indicates a sensor and the subscript *m* indicates a motor.

$Port_s$	$Starboard_s$	$Heat_s$	$Port_m$	$Starboard_m$
1	1	0	1	1
1	0	0	1	0
0	1	0	0	1
1	1	1	0	1
1	0	1	0	0
0	1	1	0	1
0	0	1	0	0
0	0	0	0	0

which, in turn, result in new sensory–motor loops. An example of a sensory–motor loop is 100–p where '100' is a sensory state indicating that the port light sensor is on (see table 9.1) and 'p' indicates a turn to port.

However, this proximal description of robot behavior is quite unlike that used by the experimenter observing *Beast₁*. Here the behavioral terms used were words like 'avoid', 'approach, 'attack', 'scurry', 'bask' etc. And the light was said to 'attract'. We call these *distal*[3] descriptions of behavior. These high-level terms describe the temporally extended behavior of *Beast₁* in its interaction with the environment. They provide subjective labels for sequences of sensory–motor loops. For example, suppose that a light is shone onto one of the light sensors of *Beast₁*, a particular sequence of sensory–motor loops may run as follows:

100–p, 100–p, 100–p, . . . ,
100–p, 110–f, 110–f, 110–f, . . . ,
110–f, 111–s, 111–s

where s is a starboard turn and f is a forward motion. The distal description of this sequence may be 'approach light' to one observer, 'attack light' to another, and maybe even 'attracted by light' to another.

Although distal descriptions are observer dependent, the idea tells us much about the emergence of behavioral complexity from simple mechanisms. As can be seen from table 9.1, the proximal description of the behavior is as simple as the control mechanism itself. However, by iterating the proximal system's short-term influence on the surroundings, the distal behavior is influenced in regular ways to generate patterns of events over longer time-scales, what the observer sees as autonomous behavior. In other words, proximal sensory–motor couplings constitute short-term self-organizing processes which over longer time-scales converge to form what we perceive to be the distal, functional regularities of behavior.

---

[3] The distal idea is a close conceptual neighbor of Von Uexkuell's *Umwelt* [320], Gibson's affordances [120], and Varela's enaction [327].

9.2.3   An artificial neural network brain

In order to get a robot to move at all we need a controller that forms a link between the sensor states and the motor states. This is the core of most robotics research. $Beast_1$ has a very basic (three-wire) artificial neural network controller. It would probably be considered ludicrous to call this a mind, even a simple one. However, it points at the foundations of the materialist mind proposed by the 17th-century philosopher Hobbes who believed that thought could be mechanized, or more profoundly, that matter could think. Much of Hobbes' thinking still persists in this century with the neo-associationists [11]. The new associationism did not take on board all of the assumptions of the early associationists, only that it is a mechanistic theory of how elementary mental elements are associated with one another to provide a causal explanation of cognition with a minimum of theoretical assumptions.

In the same spirit, the development of a formal computational approach to finding a physical basis for mind began with McCulloch and Pitts' (1943) work on 'A logical calculus of the ideas immanent in nervous activity' [194]. In this, they provided an influential computational analysis of what they believed to be a reasonable abstraction of brainlike systems. To show how thought might be computed by a brain, McCulloch and Pitts turned to the mid-19th-century work of George Boole. Boole had attempted to describe the laws governing thinking by developing a system of logic in which symbolic propositions about the world could be combined to form other statements about the world using logical connectives such as AND, OR, and NOR. What McCulloch and Pitts showed was that, by partly ignoring and partly simplifying the physical and chemical complexity of the nervous system, it was possible to build their abstract model neurons into networks capable of computing all Boolean (logical) functions. Their first simplification arose from the observation that neural communication is thresholded. That is, the spike action potential is all or none; it is either active enough to fire fully or it does not fire at all. This is similar to the threshold on the binary sensors of $Beast_1$. Thus the neuron could be conceived of as a binary computing device, an idea said to have inspired von Neumann when designing the modern digital computer. The other important simplification was that the synapses could be considered to be numerical weightings between simple binary computing elements (just like the values on the three wires of $Beast_1$). Computation proceeds by summing the weighted inputs to an element and using the binary threshold as an output function (in our case to the motors or actuators, as they are sometimes called).

We can think of the nervous system, or, more properly, the control system of $Beast_1$ as implementing two Boolean functions using two McCulloch–Pitts nets; one function for each output/motor unit. An examination of table 9.1 shows that these functions are P & H for the output to the port motor, and simply S for the output to the starboard motor. By building up large networks made up of smaller McCulloch–Pitts networks it would be possible to develop a much larger and more complex artificial neural brain capable of leading our beast through many great feats of singing and dancing. But such a controller will be brittle as we shall see.

9.2.4   Adaptivity: the beast and the hairdryer

If we are to equip a robot with a flexible (artificial) neural mind, the godlike (hardwiring) approach to the design of the neural nets will not work well. The design of complex control systems is typically a laborious effort. In many such approaches, hiding behind the success is a graduate student who spends many hours carefully designing, testing, and re-designing the set of control modules, until the desired behavior is achieved [99]. Moreover, such an approach may be too inflexible and rigid to be able to handle novel events and circumstances that the designer has not considered. Of course, swapping the wires or altering the weight values will change the association between sensors and actuators and thus the behavior, but the designer cannot chase after the robot forever.

Control systems should, in contrast, be robust to parameter variations and consist of generalizable substructures that can be re-used in order to increase, in stages, the complexity of the behavior. Neural networks offer possibilities for training the embodied control system within its own environment. In particular, real-time neural control systems that make no distinction between a learning and a performance phase are promising candidates for the control of adaptive behavior [328, 99, 317].

As an example to show the need for adaptive behavior we return to our $Beast_1$. Now $Beast_1$ has come equipped with a built in control structure that uses a heat sensor to protect it from having its sensors melted off. Such an event would leave it blind and paralysed and should be avoided. However, the mythical designer of the beast was a somewhat lazy thinker who forgot to consider attacks by non-light producing heat sources. Imagine, for example, that the beast is sitting on the floor minding its own business, i.e., it is currently not being attracted by lights. Another dark vehicle approaches that is equipped with a front mounted hair dryer and attacks our friendly $Beast_1$.

Now the trouble is that our friend, although suffering severe sensor burns, has no way to learn or adapt to such situations. The single negative wire from the heat sensor will only turn the robot if it is in the presence of a light source. As can be seen from table 9.1, the sensor states 001 will result in output motor states of 00. But of course the ever-following designer can jump in now and connect a new positive wire between the heat sensor and the starboard motor. Thus the next time its sensors are attacked by a hair dryer it will rotate anti-clockwise to escape. However, it would be better if we had a robot, let us call it $Beast_2$, that had a means of adapting to new circumstances automatically. It is to this point we now turn.

9.3   Robot training methods

Because of their adaptive nature, neural network learning is a promising candidate for implementing the control systems of autonomous robots. In section 9.2 we showed that the behavior of a simple robot can be fully described with a look-up table. It was also shown that this function can be hardwired in a very simple neural network with McCulloch–Pitts neurons. In designing more complex control structures the proximal function is, however, not known *a priori* and can be learned from examples. Rather than hardwiring the beast's network, the weight values can be adjusted by using one of a variety of learning paradigms. Thus, instead of completely engineering a control system for a specific task, a robot can be trained using incomplete data and then rely on the generalization characteristics of its neural network to adapt to novel and unpredictable changes in the world. For instance, when one of the sensors fails or when a path is blocked the robot should still succeed in its mission. The rest of this section will discuss the issues involved and present some common training methods. It will be shown that a wide variety of neural learning paradigms are currently being used in robotics. In most cases the neural control systems learn to perform a mapping between the sensory input data and the actuator states. In other cases neural networks are used to develop subsystems for a controller. Extensive references to original research are given throughout the text. Unfortunately there is not space here to do justice to the great variety of robotics research currently employing neural networks. However, we take the next best option here and provide broad coverage supported by many references which the interested reader can explore.

9.3.1   Hebbian learning

Oddly, the first proposal for neural network learning came just over 50 years before McCulloch and Pitts' famous paper. William James [152], the great philosopher and psychologist, speculated about how and when neural learning might occur in the brain. His idea was that when two processes in the brain are active at the same time, they tend to make permanent connections (e.g., the *sight* of an object and the *sound* of its name). But it was not until six years after the McCulloch–Pitts revelations that the psychologist Donald Hebb made James' learning proposal concrete. Although he cites neither James nor McCulloch and Pitts, Hebb took a step beyond them in attempting to causally relate memory and perception to the physical world [135]. His idea was that the representations of objects may be considered to be states (or patterns) of neural activity in the brain. He proposed that, each time a neural pathway is used, there is a metabolic change in the synaptic connection between the neurons in the path that facilitates subsequent signal transmission. In this way, the more often two neurons are used together, the stronger will be their strength of connection and the greater the likelihood of one activating the other. The synaptic connections come to *represent* the statistical correlates of experience. Thus in learning to recognize objects, groups of neurons are linked together to form assemblies (the neurons in any assembly may come from many areas of the brain e.g., visual and motor etc).

This notion of modifiable synapses, or synaptic plasticity, and its role in learning and memory still persists today. Although to some in the neuroscience community Hebb's ideas are over-simplistic, it has to be remembered that little was known about these issues in his day, and he did not have the technology to carry out the physiological experiments. Indeed, it was not until 1973 that Bliss and Lomo first reported, in detail, that following brief pulses of stimulation, there is a sustained increase in the amplitude of electrically evoked responses in specific neural pathways. This is the now well known *long-term potentiation* (LTP) phenomenon. Moreover, within the last ten years considerable research has shown that one of a variety of synaptic types is indeed a Hebbian synapse [9].

This is the first starting place for roboticists intent on developing physical minds for their machines. Several experiments reported in [138] use variants of the Hebbian learning rule to train light finding and avoidance behaviors on a real robot. They showed that by using this method, weights on the wires between light sensors and the motors automatically took on the desired values (+1) if a motor is switched on while the light hits the according light sensor. This resembles a 'walking the dog' method where the trainer tells the dog what action to take in what situation. After training, the dog will undertake the actions by itself. Similarly, once exposed to light in conjunction with motor commands, the robot will move towards every light source it sees.

9.3.2 Backpropagation learning

A more advanced, but related, learning method is reported in [280]. Here a Nomad200 mobile robot (shown in figure 9.2) was trained on obstacle avoidance and goal finding. In the experiments, the distance sensor and motor values are all continuously varied thus a binary look-up table such as in table 9.1 will no longer suffice. In order to learn the mapping between sensors and actuators for avoiding obstacles the robot was given a simple reflex behavior of 'move away if an object gets too close'. This treated objects approaching the sensors as increasing forces. These forces acting upon the sides of the robot (as measured by distance sensors) determined the direction in which the robot was steered e.g., steer a little left when there is a large obstacle on the starboard side. This reflex behavior was used to supervise the training of a more sophisticated neural network controller (shown in figure 9.3) as the robot blundered around its world. This is analogous to the behavior of some animals, such as giraffes, that are born with the ability to walk around the environment in a jerky motion until they learn smooth seamless behavior.

This multi-layer network was trained on the set of sensor–action pairs using

Figure 9.2. The Nomad200 mobile robot.

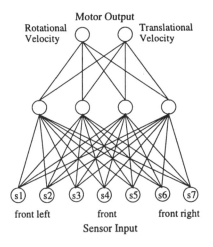

Figure 9.3. A neural control structure.

error-backpropagation learning [344]. The error-backpropagation procedure adjusts the weights between layers of nodes so that the inputs (sensor readings) of the training set are associated with the output patterns (actions) of the training set. This is a supervised learning procedure because the output patterns must be explicitly specified. During the training phase the backpropagation procedure minimizes the errors of the outputs. After training, i.e., when a certain low error level is achieved, the network is used to spontaneously produce correct output values on any input. This is called generalization performance. In robot terms, this means that the robot was able to drive through environments without collisions even when the obstacles were moved.

A similar supervised learning method was used for the goal finding behavior. Trigonometry was used to train the robot to use dead reckoning which it took from its onboard odometry. Dead reckoning is the computation of change in position by integrating velocity with respect to time. The teaching signal to the robot's network was essentially the same as someone continually pointing the robot in the direction that it should take. It is well known that such dead reckoning methods are used by a number of animals such as ants, rats, geese, and gerbils [116] but it is unclear from the animal studies how such an ability is implemented in animals. The point of the study was not to show naturalistic learning, but to show that the behavior *could be* easily implemented by a neural net. It took only 66 training examples and a short learning time (3797 learning cycles) for the robot to learn almost perfect dead reckoning behavior in a wide range of tests in the laboratory.

The goal finding and obstacle avoidance behaviors were combined on the robot in two different ways for comparison. One way that they were combined was in the form of a simple subsumption architecture [48]. The robot used the goal finding module to drive towards the specified location and switched to the obstacle avoidance module when it encountered obstacles. The other way to combine the behaviors was to train a single net on both goal finding and obstacle avoidance. A trajectory plot for the two methods of training the behaviors is shown in figure 9.4 with the subsumption architecture on the left and the combination network on the right. Although the training was on a real robot, we used the simulator to generate the plots. For more details the reader is referred to [280].

Figure 9.4. Trajectories of the robot driven by a subsumption architecture with two neural network controllers (left) and a single neural network controller (right).

Each of the plots shows a trajectory to three different goal positions along a single path. The robot starts at the lower left-hand corner and successively passes through goal locations at the upper right-hand side, the lower right-hand side and ends at the position where it is marked with a circle. It should be noted that the combined network produced smoother behavior than the subsumption architecture. This work demonstrates how general insights from animal learning, such as the use of dead reckoning and reflex behavior, can help in training adaptive robots.

9.3.3 Reinforcement learning

What the methods described in the previous sections have in common is that the teaching signals represent direct instructions to the neural network. The supervision signal gives explicit output activations for each sensor situation that occurs. This forms a so called 'bootstrap' problem similar to the story of Baron von Munchhausen who found himself stuck in the swamp. He lifted himself out of it by pulling the straps on his boots. The analogy to our robots is that they have to learn something which they should already know. In the Sharkey *et al* study, we used the idea of a simple reflex in order to bootstrap the learning. However, there are other model free methods that do not require so much in the way of built in behavior.

A commonly used approach is derived from psychological research on animal conditioning. In Pavlovian or classical conditioning [226] a conditioned stimulus (e.g., bell sound) is experimentally associated with an unconditioned stimulus (e.g., food). After a sufficient number of pairings between the conditioned and unconditioned stimulus, the conditioned stimulus alone creates the same effect as the unconditioned stimulus (e.g., salivating to the sound of a bell). This method can be used to associate a new stimulus to an existing stimulus–response behavior.

When no stimulus–response chain exists, operant conditioning gives the solution. In the early operant conditioning experiments, Skinner trained rats and pigeons to press levers to obtain food [287]. This was done by giving rewards (e.g., food) for desired behaviors and punishments (e.g., electric shocks) for undesired behaviors. The animals learned to associate their behaviors with the rewards and punishments. Neural network reinforcement learning is a methodology that attempts to capture the essence of conditioning. There are numerous examples in the literature where robots are trained using this kind of learning [195, 328, 338].

Consider a simple example in which a robot has to learn to avoid obstacles. At the start it does not have any knowledge about the relation between obstacles and sensor information and what actions to take. The robot starts with a (random) behavior determined by its randomly initialized network and the sensor signals. As a result it will explore the environment and bump into obstacles. Whenever a bump occurs, the last action is evaluated as undesired and a punishment is given. So rather than specifying the exact values to be sent to the motors the evaluator simply tells the control structure whether the action was right or wrong. This signal then has to be used by the learning rule so that the number of future punishments is minimized. If this is applied in the right way, the obstacle avoidance behavior of the robot improves by experience.

In an obstacle avoidance task, it is relatively easy to detect undesired actions such as bumps. An external observer actually sees or hears the robot bump or when the robot is equipped with whiskers the obstacles can be detected by the robot itself. In the latter case the evaluation function (or heuristic) can be implemented as an 'instinct' rule allowing the robot to learn completely autonomously. The exact definition of the evaluation function determines the resulting distal behavior but for complex tasks it is not always a straightforward job to define such an evaluative function. When the evaluation function gives both rewards (for desired actions) and punishments (for undesired actions) the developing control structure must find the right trade-off between exploring and exploiting. In the beginning, rewards can only be received by exploring at the cost of punishments. But the number of rewards should gradually increase when experience grows and the number of punishments decreases. The robot should thus change its behavior over time and move from exploring towards exploiting. An extensive overview of the explore and exploit dilemma is given in [333].

Besides designing a suitable evaluation function, a learning rule must be implemented for adjusting the weights. In the Hebbian and backpropagation learning paradigms as described above, the desired node outputs were given explicitly by the instructor and the implementation of a learning rule was rather straightforward. However, in the reinforcement learning paradigm the desired node outputs are unknown and must be derived from the evaluation function. A method that works well is described in [195] and is called complementary reinforcement backpropagation (CRBP). The spontaneous outputs of the initial network are taken as probabilities that are used to stochastically produce some output action (e.g., move forwards, turn left). If a reward is received for this action the network will be 'pushed' into the direction of the actual (desired) output. If a punishment is received the network is pushed into the complementary value of the current output. Experiments have indicated that these implementations learn even faster when some noise is added to the sensor values. This helps the exploration function in covering the search space.

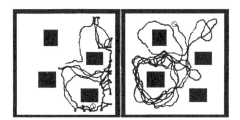

Figure 9.5. From bumping behavior (left diagram) to avoidance behavior (right diagram) by reinforcement learning.

Figure 9.5 gives a trajectory of the first 1000 learning steps of a robot trained with CRBP reinforcement learning with a feedforward neural network.

The robot's trace (diagram on the left) shows where the walls and obstacles were touched. The robot was turned 180 degrees whenever that happened (22 occurrences). The diagram at the right-hand side shows the behavior of the robot after 80 000 training steps. The robot now shows perfect avoidance behavior with no collisions during 1000 steps.

Reinforcement learning on real robots is slow because the robot learns through repeated trials that are limited by the physical speed. Another problem with real robots is that no dangerous trials can be used since these could lead to severe damage. A method to overcome both problems is proposed by Nilsson [216] who equips his robot with a self-simulator that allows the robot to 'imagine' itself interacting with the environment in various hypothetical simulations. Nilsson calls this method 'robot consciousness' because the robot holds and manipulates a model of itself. Provided that the self-simulator is fast and accurate enough, many trials can be done without physical execution and so the learning process is speeded up. In addition, harmful trials can easily be backtracked, preventing the real robot from damaging itself. An implementation of this method is presented in [217] where a snakelike robot is trained to move in a desired direction. The dynamics of this robot are simply too hard to be grasped by a human programmer. The proposed self-learning method proves viable to get the robot to crawl effectively in real time.

9.3.4 Evolutionary learning

The evaluation functions used in reinforcement learning are all used to adjust weight values of an existing control structure. This type of learning is called ontogenetic. Evaluation functions can also be applied to longer time-scales than the *life-time* of a robot. In this phylogenetic or evolutionary learning the performance of the control structure as a whole is evaluated. Rather than developing one neural control structure by tuning the weights, in the evolutionary learning approach complete generations of control structures are evolved. The following early example of evolutionary learning applied to robots with neural control structures is taken from Valentino Braitenberg's book [45].

Imagine you have a table with sources of light, sound, smell and landmarks indicating the edge of the table. On the table are a few robots driving about. You are supplied with a tool set including batteries, threshold devices, wheels, motors, sensors, wires, tin and so forth. What you then do is take one of the robots that is running on the table top and build a copy of it. Both the model and the copy are set back on the table and you pick up another one. You ignore robots that fall off because they are not able to cope with their environment. You repeat this process and try to produce robots at a pace that roughly matches the rate at which robots fall off the table. Because you are in quite a hurry it is likely that you introduce errors during the copy process. This might result in a useless new robot that drives off the table as soon as it is put there. On the other hand, the change in wiring can also improve the robot's behavior. The new robot then 'survives' much longer than its parent and is more likely to be picked up and be used as a model for new ones. You will experience that after a while your work load diminishes as the robot colony on the table is adapted to survive in the environment. This is all the result of a Darwinian evolutionary process in which generations of robots were reproduced according to rules of selection!

This is an example of evolutionary learning in hardware, i.e., the hardware itself is changed. Currently, only a few researchers are experimenting with evolution in real robots because it is quite costly both in development time and hardware resources. Much more research is dedicated to evolving neural controllers in simulation that can either be used in simulated robots or real robots [189]. The method used in these studies is referred to as genetic programming (GP). In GP, evolution starts off with a number of genotypes (first generation). Then the best-behaving genotypes are selected on the basis of a fitness function (e.g., robots that stayed on the table are fitter the those that fell off). The genetic encoding of their control structures forms a basis for the making of new (second generation) species. This can be done using genetic algorithm operators such as crossover and mutation. In crossover, genetic material from several 'parents' is combined in order to create the 'child'. This gives the child the possibility to inherit characteristics from both parents. The mutation process allows new characteristics to develop and can be compared to the copying errors in the example taken from Braitenberg.

Nolfi and Parisi [219] present evolutionary robot experiments where a neural control structure is evolved for a Khepera robot. The Khepera is a miniature robot that operates in an arena on a desk top, see figure 9.6. In the experiments described by Nolfi the robot is equipped with distance sensors and a gripper. The task was to 'clean' the environment by throwing small objects outside the arena. This task implies subtasks such as moving around in the arena, locating an object, picking it up, moving towards the walls, and releasing it outside the arena. The initial control structures were 100 feedforward networks with random connections. The robot behaviors with these control structures were then tested in the real world (a simulated world was used as well in order to speed up the process). From these 100 initial control structures only the 20 best performing were selected. These were used to generate five children each so that the next generation again consisted of 100 members. This process was repeated 300 times. After 300 generations individual control structures were evolved that performed the cleaning task rather well.

These experiments highlight the interesting properties of evolutionary learning. There are, however, also a number of drawbacks. For example, when the entire population of control structure generations is tested on real robots the process is very slow. Another drawback is that no designer understands how the evolved structures work. This might not seem a big problem since experiments have yet only indicated that the evolutionary learning approach is capable of evolving structures that could also be designed by hand [189]. But as soon as

Figure 9.6. The Khepera miniature robot.

the approach becomes more successful and yields more complex structures this lack of understanding will certainly cause a number of problems.

9.3.5 Beyond reactive control

In all of the examples listed above, neural networks were used to directly control the robots. These networks immediately map their sensor inputs onto the motor outputs. This means that the robot action is a direct result of the sensory input and each specific input will always lead to the same output action. Robots with such behavior are called *reactive* [14] and are very limited in their repertoire of behaviors. Simple reactive robots can even quite easily get stuck. For instance when the robot approaches a symmetrical corner, the control structure will not be able to tell in which way the robot should be steered to free itself. For example, when the robot steers left, the sensor at the left side will sense even less free space and the next best action will be to steer right. This process will alternate leaving less free space at both sides until the robot finds itself stuck in the corner! This reactive stupidity can possibly be overcome by adding a simple memory or increasing the proximal sensor action loop along the time dimension. The robot can then for instance first recognize the corner and then initiate a 'free from corner' routine e.g., by backing up or making a U-turn until it finds itself in free space again.

Various neural network paradigms incorporate memory themselves by using recurrent links that feed back information. For instance in the Jordan network [153] output information is fed back to state units that are presented together with the new input. The original Jordan network was created to generate sequences of outputs. In the robot domain it could be used for the generation of trajectories. In Elman's simple recurrent network (SRN, [346]) hidden unit states are fed back to context units. These hold some history of the former sensor action states and seem useful in recognizing certain (critical) situations. Both the Jordan and Elman networks and their variations are used in robotics. For instance Meeden *et al* [195] report experiments that show much better reinforcement learning results for an avoid obstacle task when using feedback loops than when using standard feedforward networks with no feedback loops. Figure 9.7 shows an example of a neural network with output recurrency.

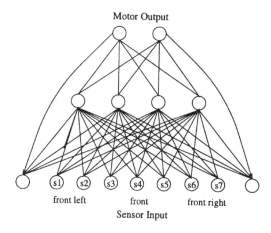

Figure 9.7. Example of a neural network with output recurrency.

The proximal sensor–action loop discussed above can also be extended by implementing a hierarchical control structure. An example of a robot control system that utilizes a number of dynamical neural networks for learning and performance at several stages is used in the mobile adaptive visual navigator (MAVIN) [29]. MAVIN learns various motor behaviors and is required to associate these with learned objects. The building blocks for visual object learning, behavioral conditioning and so on are based on the neural paradigm ART-1 (adaptive resonance theory, [345]). The actions of robots with control structures such as MAVIN are no longer just the result of the current sensor state but in addition depend on the robot's internal state that keeps a little history of the former actions.

Some neural network paradigms offer properties that are useful for specific parts of the control task. For instance self-organizing networks such as the Kohonen self-organizing maps (SOMs, [347]) find application in the pre-processing of sensory information. When exposed to large data sets SOMs will organize themselves in such a way that similarities in the data are detected by the network. Data sharing similar features are clustered on neural units that are ordered so that neighboring units represent similar clusters. This method is often used for down scaling the dimensionality of large data sets. SOMs applied to high-dimensional sensory data will yield a map that represents features of the world as it is sensed by the robot. Nehmzow [206] reports experiments where a robot explores its environment while constructing an internal representation of its world with a SOM. The robot thus does not use any knowledge from supervisors or built-in programs. After the initial self-organization phase, the internal representations are successfully used for tasks such as localization, and decision making based on visual data. Similar use of SOMs is reported in [317] where it is shown that nearby nodes in the SOM represent 'similar' real-world features such as certain obstacles.

9.3.6   A summary of neural networks in self-learning robots

We have seen from the previous sections that neural networks find application in self-learning robots in a variety of ways. A coarse distinction was made between neural networks that take complete control of the robot and neural networks that only form a part of the control system. This is mainly a result of the differences between the neural paradigms themselves. For instance, Kohonen's self-organizing maps find application in the pre-processing of raw sensor data because of the feature detection characteristics. Modular networks such as ART call for implementation in hierarchical structures that consists of many processing levels. Other neural paradigms such as backpropagation are, however, used in all kinds of robot control structure. The category of complete neural control structures can be subdivided into instructive and evaluative training methods. Instructive training methods require explicit instructions. They need a supervisor that dictates exactly what actions are to be taken at every time step. The evaluative training methods, in contrast, do not need such precise instructions. An internal or external supervisor can be used to judge the behavior at a higher level by choosing between right and wrong actions. When these actions are judged on a short time-scale, i.e., in the order of the small time steps, we speak of reinforcement learning. As was shown, many neural paradigms are used in this context. In the evolutionary approach, judgments are only made at the end of the robot's life-time or before reproduction. Many neural paradigms can be used in conjunction for the implementation of these types of evaluative learning.

9.4   Back to the future

It should be clear by now that the autonomous robotics of modern science are very far removed, as yet, from those depicted in science fiction. If anything, one could make the argument that the simple beasts that we have been discussing, including the learning robots, are in some sense, a backward step from where artificial intelligence (AI) was 20 years ago. AI developed rapidly as a field in which large-scale symbolic programs were used for tasks that would be considered to require intelligence if carried out by humans (e.g., chess playing, problem solving, language processing, and so forth).

Early AI-driven robots took advantage of having large amounts of knowledge programmed into them by designers who knew the domain. These designers programmed both the proximal and distal behavior of their robots. For example the SRI International robot, SHAKEY (see [218]) moved about in a carefully engineered world. Analyzed images from its onboard camera were used by a first-order predicate calculus model of the world before the STRIPS planning system generated sequences of action. However, such robots seemed to have reached a ceiling in terms of what they could do. It turns out that it is difficult to design systems that link models of the high-level aspects of intelligence with the real world (cf. [49] for a critique).

One of the problems with such monolithic systems, many of us believe, is that they are really dealing with only a small part of intelligence, the peak of human reasoning abilities, without considering the very much larger part of the intelligence we share with lower animals and which has evolved over millions of years of adapting to the environment. In contrast, the bottom-up robots we have been discussing start off with very primitive intelligence (or none) and have to discover the world for themselves. They bumble about semi-autonomously bumping into objects and learning from their experiences. Currently, their learning is at best crude, based on mappings between sensors and actuators to develop sensory–motor loops. Why then should we take such devices seriously? What do they have to offer us for the future? And most importantly, what to they have to do with helping us to develop simple minds on robots?

The new science of intelligence, *à la* self-learning robotics, is grounded firmly in the world; it is the study of a situated intelligence, rather than the study of *in vacuo* intelligence exhibited by most AI systems. The traditional approach to cognition (or thought) that underlies much of AI is that the mind is essentially like a piece of software running on the hardware of the brain. What is more, the mind program would just be happy running on any computational device that was capable of housing it. Meaning then consists simply of the syntactic relationship between all of the mental elements. But what of the world?, you may ask. The physical world, according to some cognitivists (e.g., Pylyshyn [245]) is related to the mind by transducers that transform light into symbols that can then bind onto variables in the mental program. One of the staunchest critics of the AI position, the philosopher John Searle, puts this view of meaning into perspective in his famous Chinese Room argument which goes roughly like this. A person, who does not speak any Chinese, is sealed in a room and passed pieces of paper through the letter box. On each piece of paper are some Chinese questions which our person answers by using a Chinese look-up table (or whatever). Then he passes the paper with the answers out through a letter box on the other side of the room. This is similar to performing syntactic transformations on the data. Searle's point is that the whole operation has no meaning to the man in the room although all of the answers to the questions may be correct.

One way to address this problem of how to give symbols meaning and yet stick within the traditional AI/cognitivism approach is to find a way to 'hook' the elements of the physical world onto symbols in the mind (not unlike Pylyshyn's transducer or indeed referential semantics (see [282]). This is referred to as the symbol grounding problem [129]. However, the approach of embodied cognition maintains that the mind or the cognitive system arises as a direct result of the *interaction* between organism and environment. It is not simply that the mind has pointers to the world; it is the interplay between beings and their world that creates minds. For example the organism must find a way to classify its own distal behavior.

Although there have been a number of worthwhile modern texts on the relationship between the new approach to mind and robotics, e.g., [327], it is to the earlier work of the biologist/philosopher/developmental psychologist Jean Piaget that we turn for inspiration.

9.4.1 Piaget and the emergence of intelligence

In his genetic epistemology, Piaget (e.g., [231]) developed a theory of how humans develop from immature biological organisms to abstract reasoning beings through a combination of sensory reflexes, innate non-domain specific processes, and interaction with an environment. The importance of this for us is in his explanation of how cognitive structures gradually emerge from the accommodation of simple built-in reflexes and some built-in modes of functioning.

Piaget argued that there were two forms of heredity: specific and general (e.g., [232]). The specific were a species-specific inheritance that conditions what we can directly perceive. In robot terms this is the basic hardware and the type of sensory equipment such as sonar or infrared. These could impede or facilitate intellectual functioning but could not account for it. As far as Piaget was concerned, it was *general* heredity that accounted for intellectual functioning and, indeed, biological functioning in general. He argued that cognitive or mental structures only come into being in the course of development. What we inherit are the functional invariants, *organization* and *adaption*, to provide us with *modus operandi* that allow us to interact with the environment in particular ways.[4] Organization refers to the internal intellectual structures and its specific characteristics differ considerably from one developmental stage to another. However, our focus here is on adaption. The use of the term adaption is very specific:

---

[4] In recent years Piaget's anti-nativist stance has been challenged by the findings of many developmental psychological findings. It is believed now that the child has many domain-specific inheritances. However, despite this, the Piagetian principles may still be reconciled with a weak form of nativism (see [157] for a review).

... if everyone recognizes that everything in intellectual development consists of adaption, the vagueness of this concept can only be deplored... It is therefore necessary to distinguish between the state of adaption and the process of adaption. In the state, nothing is clear. In following the process, things are cleared up. There is adaption when the organism is transformed by the environment and when this variation results in an increase in the interchanges between the environment and itself which are favorable to its preservation. ([232], p 5)

Equilibrium in the biological sense was a favorite term of Piaget. Adaption was seen as an equilibrium between the action of the organism on the environment and vice versa.

Adaption is made up of the twin processes of *assimilation* and *accommodation*. In brief, assimilation:

... may be used to describe the action of the organism on surrounding objects, in so far as this action depends on previous behavior involving the same or similar objects. ([231], p 9)

In terms of our robot examples, when a robot has learned to avoid boxes we would expect it to assimilate walls and cylinders into its avoidance activity. Accommodation, on the other hand, is understood as a change in the organism brought on by pressure of circumstances in the environment. For example, if our robot travels into a novel environment filled with new obstacles, such as chairs and tables or star shaped aspects, it may begin by bumping into them causing changes (through, for example, reinforcement learning) in its neural net controller.

Perhaps the most famous aspect of Piaget's work, mainly because of its accessibility, is his idea that on the route to adulthood, children pass through four invariant stages of development:

(i)   Sensory–motor.
(ii) Pre-operational.
(iii) Concrete operations.
(iv) Formal operations.

Mostly, autonomous robotics could crudely be classified at the lower end of the sensory–motor stage with some ambitions towards the pre-operational stages. AI on the other hand would prefer to be more linked with the final two stages of concrete and formal operations.

Looking back at section 9.3 on training methods for robots, it should be clear that we can derive some analogies to the work of various components of Piaget's sensory–motor stage. However, no one has 'bitten the bullet' yet and tried to design an autonomous robot that actually develops in the way described by the Piagetian program.[5] Yet it is inevitable that such a program will eventually

---

[5] But see Bakker and Kuniyoshi [25] who have plans in this direction.

have to be invoked if our robots are going to exhibit more intelligence than the average vegetable. To understand such a program we need to take a brief tour through the six sub-stages of the emergence of sensory–motor intelligence and then touch on the emergence of the pre-operational mind (cf. [52]).

The sensory–motor stage: 0 to 2 years

- **0–1 month:** The infant reacts by a series of innate reflexes and accommodation modifies the preparatory activity so that the nipple can be more easily located. As we pointed out earlier, the existence of simple reflexes are a necessary condition for the emergence of intelligence. In an excellent book interpreting Piaget's works, Flavell [104] writes:

  > ... the simple reflexes with which the neonate is endowed soon undergo definite modifications as a consequence of environmental contact; in so doing, they imperceptibly become acquired adaptions instead of mere reflexes— 'wired-in' responses of purely endogenous determination. Thus birth reflexes are truly the building blocks of the sensory–motor edifice; intelligence begins with them and is constituted as a function of their adaptation to the environment.

- **1–4 months:** The infant systematically sucks its thumb through a new learned behavioral pattern. A series of actions that were initiated by a change in response leads to the desired outcome (a thumb in the mouth) and tends to be repeated through assimilation. This is described by the term *circular reaction*. In this sensory–motor substage, the reactions are focused on the infant's body and not on the environment and are thus called *primary circular reactions*.

- **4–8 months:** *Secondary circular reactions* emerge which modify motor habits rather than reflexes (as in the previous sub-stage). This is because the reactions are now focused on the infant's environment as well as the body. This stage is characterized by a form of 'motor recognition'. That is, the infant may show recognition of an object by performing a version of the usual motor activity associated with the object. There are also some signs of primitive intentionality in that after some fortuitous behavior produces an interesting effect, that behavior may become a goal for secondary circular reactions.

- **8–12 months:** *Secondary schemata* emerge from the coordination of secondary circular reactions. Essentially these schemata are stored sequences of sensory–motor loops. This stage is characterized by:

  — Indications of intention.
  — Anticipatory behavior (but no sophisticated imagery, what there is still bound to infants' sensory–motor behavior).
  — Object permanence is extended by superior manipulation skills.

- **12–18 months:** The *tertiary circular reactions* emerge. Unlike the secondary circular reactions, these are not characterized by stereotypical behaviors. Rather, tertiary circular reactions are more of an exploration of the achievement of the goal.
- **18–24 months:** The main change here is in the appearance of rudimentary mental imagery which give rise to the beginnings of internal representational states. Now sequences of exploratory behaviors in the previous stage become more covert. The behaviors can be simulated internally before an action is carried out.

9.4.2 A new representationalism?

What we have seen of the Piagetian sensory–motor period is a steady progression of development from simple reflexes through a series of ever increasing abilities to perform circular reactions. Beginning with the primary circular reactions which result in the repetition of serendipitous actions related to the preparation of the reflexes, the child develops secondary circular reactions which show the emergence of goal-driven behavior. After that, the tertiary circular reactions develop as an exploration of goal achievements. All of this is leading the child towards an internalization of the world so that it can think without simply reacting and can simulate what might happen under different circumstances. Thus it is necessary to develop representations with increasing flexibility. This is preparation for the *pre-operational* stage which is characterized by the extensive development of mental imagery. At this stage, the infant is capable of imitating the actions of others and this may be deferred until the model is not present. There is also promise in some recent work on imitation which should take us on to a stage beyond sensory–motor intelligence (see [25] for a review). This work is just beginning and has a long way to go, but it shows a way forward out of simple reactive behavior. At present the tasks are behaviorally very simple such as following a robot teacher through a maze [134] or over a hilly landscape [80] where the actions to be copied are restricted to a very small set such as *move forward* or *turn right*. Kuniyoshi *et al* also used a small number of actions but tackled the problem of recognizing a human teacher on an assembly task [166]. At present, the research does not satisfy the necessary developmental conditions from the Piagetian viewpoint. Firstly, there is no development through the sensory–motor stage to allow the internalization of the world. Secondly, the behaviors are always imitated explicitly rather than covertly. Finally, the representations are all pre-determined rather than developed. Indeed, this underutilization of learned representations is common in the robotics literature.

A problem with the new connectionist representations is that, like the old, they have mainly been taken from training *in vacuo* rather than being taken from training in the world. It has been argued that at least, unlike good old-fashioned AI, neural network representations are developed by extensional programming, i.e., from examples. However, in developing naturalistic intelligence the learning programs really need to be connected to the physical world.

One of our current major research goals is to link situated robotics to this new representational form and to develop representations from interactions with the world. Probably the best route, at present, is to take up the Piagetian program and develop the representations incrementally. This will help to loosen the environmental puppeteer's strings. For example, when discussing behavioral descriptions of robots in section 9.3, it was pointed out that the moment to moment movement of the robot could be described in proximal terms whereas the long-term trajectories are described in observer-dependent distal terms e.g., avoidance. However, for a robot to move beyond its current perspective of reacting to its world it must also become an observer of its own behavior. That is, it needs to develop internal representations that will help it to classify its own distal behavior at greater and greater temporal–spatial distances and have the ability to use these representations. We believe this to be a very promising approach for autonomous robotics. The problem is that it may be a very slow approach for the world of fast-paced academic products. Nonetheless, it may in the long run be the most worthwhile if we are to begin to develop a neural mind for a robot.

9.5   Conclusions

We started this chapter by discussing the myth and reality of robotics research. Moving from the magic and fiction of the powerful public conception of robots, it is difficult to make an impact on the imagination with what essentially looks like a dustbin on wheels (see figure 9.2). However, our hope is that we have convinced you that the study of self-learning autonomous robotics is a worthwhile scientific venture. If for no other reason, it makes us think more clearly about what we mean by autonomy and how this concept relates to real life. It should also increase our understanding of the animals that we model. But most importantly for us, it provides us with a platform for research on the construction of artificial beings whose intelligence emerges both from their interplay with the environment and the gradual development of representations of their actions in the world.

Whether we develop a Pinocchio or a Terminator only time will tell. In the meantime it is essential that we avoid the *spin doctoring* often associated with artificial intelligence. Our noses have not grown too much in writing this chapter.

Acknowledgment

We would like to thank the Engineering and Physical Sciences Research Council (UK) for finding this research (grant number GR/K18986).

Chapter 10

Teaching a Robot to See How it Moves

Patrick van der Smagt
German Aerospace Research Establishment, DLR, Germany
smagt@dlr.de

10.1 Introduction

The positioning of a robot hand in order to grasp an object is a problem fundamental to robotics. The task we want to perform can be described as follows: given a visual scene the robot arm must reach an indicated point in that visual scene. This point indicates the observed object that has to be grasped. In order to accomplish this task, a mapping from the visual scene to the corresponding robot joint values must be available. The task set out in this chapter is to design a self-learning controller that constructs that mapping without knowledge of the geometry of the camera–robot system.

When the position of the object is unequivocally known with respect to the robot's base, and the robot geometry (the kinematics) is known, a single computation followed by a robot joint rotation suffices to reach the indicated position.

But what if these data are unavailable or, in a more typical case, are too inaccurate to solve the problem and grasp the object? To tackle this problem, we assume an academic problem and discard any explicit model of the robot or its sensors. We assume that the visual system needs no precise calibration. The solution is given by *learning*: a neural controller has to learn to generate robot joint rotations which position the end-effector directly above the target object.

Figure 10.1 demonstrates the task. Note that, due to the fact that the designed system does not rely on a model of the robot, it can in fact be applied to *any* robot arm.

In order to design a system which can be successfully used in real-world applications, there are two important issues which have to be considered. Firstly, in real-world applications of robot systems, a 'reasonable' training time must be

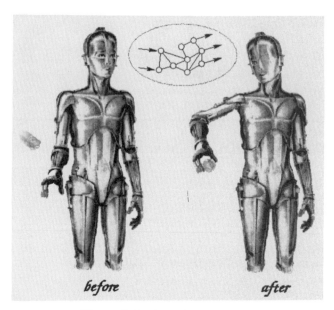

before   *after*

Figure 10.1. The task: grasp an object.

ensured. Real robots move slowly in comparison with simulated robots, so it is important that after only a few trials the goal is reached. Hundreds of trials are not acceptable. Secondly, the added value of self-learning systems must be fully exploited: it is essential that the method adapt to unforeseen gradual or sudden changes in the robot–camera system. The combination of these two points has been ignored in many previous approaches.

The system is thus set up so that the relation between sensor input and robot motion is many-to-one, i.e. given a sensor reading, a robot motion can be *uniquely* determined. This simplification means that there is always only one posture (arm configuration) to reach a specific position; for instance, there is no elbow up/elbow down ambiguity. As customary in robotics, this situation is realized by choosing a preferred situation (e.g. the elbow up configuration) and never presenting the conflicting (elbow down configuration) to the adaptive controller. Note that it is not guaranteed that the elbow down configuration will never occur; when the controller parameters are sufficiently disturbed, this case is, in theory, not excluded. In practice, however, it is never learned and therefore does not occur.

In our approach no explicit models of the camera or the robot are available. The camera–hand mapping must be learned by the neural network based on the (measured) behaviour of the camera–robot system. *Learning samples* are gathered during the control process and added to the *learning set*. The size and exact implementation of this learning set are directly related to the adaptability and accuracy of the system: a large learning set leads to a sluggishly adapting system, whereas a small learning set cannot be used to construct an accurate controller. Therefore its size must be variable.

Summing up, we have to tackle the following four problems:

(i) How do we control the robot without having a model of the robot or of its sensors?
(ii) How do we get the (computationally intensive) neural network to learn the robot's behaviour on-line?
(iii) How do we ensure that the resultant approximation is precise enough to control the robot in its large reach space?
(iv) Is the neural network fast enough for real-time robot control while handling visual data?

To understand what these requirements mean we first need to understand some of the pitfalls of robotics and vision.

10.2   The components

10.2.1   Robotics

In robotics, when we restrict ourselves to the control of robot arms, we are faced by three problems:

(i) Assuming that the target position (in Cartesian or sensor space) to which the hand of the robot arm must go is known, a set of joint angles must be computed with which the robot can reach that position. This problem is known as *inverse kinematics*.
(ii) Secondly, a path must be generated along which each of the joints must be moved in order to reach the target position from the current position. This problem is known as *path planning*.
(iii) Finally, the correct forces must be exerted on the joints (e.g. by giving the motors of the robot the appropriate currents) in order to actualize the motion. This problem is known as *inverse dynamics*.

Computation of the inverse kinematics is a solvable problem, provided that the dimensions of the robot are known. This knowledge depends on a model of that robot, and of course a problem exists when a model is not available or not very accurate (which is often the case). One should realize that an error of a fraction of a degree in robot arm rotation can result in from a few millimetres up to a centimetre of positioning error of the robot hand. The model therefore has to be very precise.

More complex are the inverse dynamics. The dynamics of any d-degrees-of-freedom robot with rotational joints can be described by the equation [75]

$$T\left(\theta, \dot{\theta}, \ddot{\theta}\right) = F_1(\theta)\ddot{\theta} + F_2(\theta)\left[\dot{\theta} \cdot \dot{\theta}\right] + F_3(\theta)\left[\dot{\theta}^2\right] + F_4(\theta, \dot{\theta}) + F_5(\theta) \quad (10.1)$$

where T is a d-vector of torques exerted by the links, and $\boldsymbol{\theta}$, $\dot{\boldsymbol{\theta}}$, and $\ddot{\boldsymbol{\theta}}$ are d-vectors denoting the positions, velocities, and accelerations of the d joints.

$$[\dot{\boldsymbol{\theta}} \cdot \dot{\boldsymbol{\theta}}] = [\dot{\theta}_1\dot{\theta}_2, \dot{\theta}_1\dot{\theta}_3, \ldots, \dot{\theta}_{d-1}\dot{\theta}_d]^T \qquad [\dot{\boldsymbol{\theta}}^2] = [\dot{\theta}_1^2, \dot{\theta}_2^2, \ldots, \dot{\theta}_d^2] \qquad (10.2)$$

$F_1(\boldsymbol{\theta})$ is the matrix of inertia, $F_2(\boldsymbol{\theta})$ is the matrix of Coriolis coefficients, $F_3(\boldsymbol{\theta})$ is the matrix of centrifugal coefficients, $F_4(\boldsymbol{\theta}, \dot{\boldsymbol{\theta}})$ is a friction term, and $F_5(\boldsymbol{\theta})$ is the gravity working on the joints.

When the robot has to move from one joint position to another, a torque must be applied which generates T. The problem of calculating the correct torques (forces) to have the robot arm follow a specified trajectory is known as *inverse dynamics*. Industrial robots are generally designed to eliminate the interdependence between the joints, such that the robot arm can be regarded as d independent moving bodies. This reduces the vector field described above to d independent functions of three variables for which the coefficients have to be found. In addition the link actuators are usually made so powerful that F_1, F_3, F_4, and F_5 can be considered independent of $\boldsymbol{\theta}$. For this simplified (and common) case, various standard methods exist to compute the inverse dynamics [113]. This controller eliminates the requirement of knowledge of the robot arm in order to control it. When using such a control method, the robot arm can be controlled by specifying joint values, velocities and accelerations, and knowledge of the required forces is not required. However, this control method requires a precise model of the robot, which may not be available. Adaptation and learning are therefore important tools.

10.2.2 Vision

The vision problem is different: here we are faced with huge numbers of data (in particular, a black-and-white camera generates approximately 0.25 Mb of data each 40 ms, i.e. over 6 Mb of data per second, a colour camera three times as much). How are we going to handle these data?

Clearly, even the fastest workstation, which can make over 100 Mflops per second, has no chance. Although six million multiplications can be done in about 0.4 s, if in the meantime six million memory locations have to be accessed this time can go up to more than 10 s.

Therefore it is wise to pre-process the visual data. With specialized hardware, we might do the following tricks in real time (i.e. as fast as the images are generated):

- Reduce a grey-level image to black and white (two-colour);
- Sub-sample the black and white image, i.e. reduce a full-size 512×512 image to 256×256 or 64×64. In fact, reduction up to 32×32 is possible without losing data; the binary black and white image needs only one bit per pixel.

- Select all the regions which are white and compute their position and orientation in the image (blob finding and moment computation).
- Select the object that we are really interested in, and find a unique form for its position (three integer values) and orientation (three floating point values).

Thus the 0.25 Mb per 40 ms can be reduced to a few bytes only. These data can be easily processed by a neural network or any other controller.

Nice as it seems, there is of course a price to be paid. The flexibility of the resultant system is limited due to the approach. As much as we would have liked to, it is just not feasible to pump the whole image into a self-learning neural system!

Naturally, there are exceptions to this rule. One of the most famous examples is the ALVINN, a neural-network-based vehicle driving system. The neural network has as input a sub-sampled 30×32 pixel image from a camera mounted on the roof of a car. The input is fully connected to a five-unit hidden layer, which in turn is connected to 30 output units, each of which indicates a direction in which the vehicle has to steer. To teach ALVINN to steer, a driver has to drive the car for about three minutes while ALVINN is learning. Due to its generalizing ability, ALVINN is not only able to drive on roads marked with white and yellow stripes, but has demonstrated its ability to stay on single-lane dirt roads, single-lane paved bike paths, two-lane suburban neighbourhood streets, and lined divided highways. On the last domain speeds of up to 70 mph were reached on public highways. Note, however, that for each road type a network must be specifically trained. The choice of which network is best used is also taken care of by ALVINN.

10.2.2.1 Visual setup

For the system to accomplish the specified task it must know the position of the observed object relative to the robot, in *some* coordinate system. Consider a robot where a specific moment has a joint position θ. The position of the end-effector in world coordinates is given by x_r. The robot has to move towards an object located at world position x_o, i.e. assume a θ such that the x_r equals x_o. As discussed above, x_o and x_r are not available without an accurate model of the visual system. There are two basic visual setups to obtain the required visual information:

(i) *External 'world-based' camera.* Both the robot and target object are observed by cameras situated at fixed and unchanging positions. The visually observed object features (indicated by $\{\xi_i\}_o$) in the image must uniquely determine the position x_o of the target object; otherwise, when the target position is not uniquely known, the required robot move to reach the target cannot be determined from the observation. Also, the visually observed robot features $\{\xi_i\}_r$ representing the robot end-effector must uniquely determine its world coordinates x_r, to ensure that the relation between the vision domain and the robot domain is learnable.

(ii) *Internal 'robot-based' camera.* The target object is observed by cameras which move together with the robot's end-effector. The observed object position (the visual observation ξ) may not uniquely describe the object position in world coordinates. However, now measurements ξ together with the robot position θ must uniquely define the world coordinates x_o if a non-ambiguous motion plan has to be made.

These two setups are basically different in the following way. In the case of world-based vision, the target object, the robot end-effector, and the (positional) relation between the two must be determined. In robot-based vision the target object is observed in a camera coordinate frame relative to the robot end-effector position. Therefore it is not necessary that the position of the robot end-effector be observed, and hence robot-based vision is simpler and more robust.

Robot-based vision has another advantage over world-based vision. The positional precision that can be extracted from a quantized camera image is inversely proportional to the distance between the camera and the observed scene (when the focus of the optical system is fixed). Thus the visual precision increases when the target object is approached; the finite resolution of the camera is no limiting factor. World-based cameras, which are not moving with the gripper and yet have to see the whole work space, have to be placed rather far away, typically in the order of 2 m from the robot base for a robot with a typical arm length of 1 m. Thus their precision will be limited, typically 0.5–1 cm for the application described in this chapter.

The perspective transformation that maps 3D points on a plane (the image plane) is a many-to-one relationship and is thus not invertible. A single image obtained from one camera does not provide the required depth information to reconstruct the positions of the observed components in the 3D world. Since the controllers developed within the scope of this thesis need such information, additional *a priori* knowledge is required. Two solutions are considered.

(i) *Model-based monocular.* In this case, *a priori* knowledge of the observed object is assumed. For instance, when a sufficient set of point features of the object can be observed, and the position of these features on the object is known, the position of the object relative to the camera in world space can be reconstructed. For more detailed information about robust model-based approaches, consult [117]. An exemplar method is the following: when the camera is looking down to the scene consisting of a single object with a flat, horizontal surface, the observed area of the object is inversely proportional to the square root of the distance.

(ii) *Correspondence-based binocular.* When no *a priori* knowledge is present, triangulation can be used to measure depth. This can be realized with a stereoscopic system [28]. When two cameras, whose image planes are situated at known (relative) positions, observe the 3D scene, the corresponding point features from both images can be used to reconstruct the 3D image. As shown in [258], the relative image plane positions need not be calibrated but can be incorporated in the learning mechanism.

When the visual scene becomes increasingly complex, or the number of degrees of freedom for positioning the robot manipulator increases (e.g. not only the position but also the orientation of the hand is of importance), the complexity of stereo vision will increase considerably: especially for complex visual scenes, the *correspondence* or *matching* problem [187].

10.2.2.2   Implementation

In conclusion, the advantage of model-based monocular vision is the increasing precision, avoidance of the correspondence problem, and simpler image processing. An additional advantage is that the problem of occlusion of marker points or parts of the observed object (e.g. due to the rotating robot arm) is avoided. A disadvantage of this method is the requirement of *a priori* knowledge of the observed scene.

Relative depth information can be obtained by using sequences of visual images. By measuring the divergence in an image when approaching an object (e.g. how much an observed object gets larger when it is approached) , i.e. the visual distance divided by the visual velocity. Note that the absolute distance cannot be measured, a large object far away cannot be distinguished from a nearby small object.

In most CCD cameras, the image plane (CCD) is placed at the focal distance f from the lens, such that the point of focus is at infinity. An object placed at distance $d_z + f$ from the lens will be projected on a point h from the lens, such that [136]

$$\frac{1}{d_z + f} + \frac{1}{h} = \frac{1}{f} \tag{10.3}$$

as depicted in figure 10.2.

In the proposed system, the depth d_z will be derived from the projected (or observed) area ξ_A in relation to the 'real' area A of the object; A is defined as the projected area of the object when $d_z = f$. This observed area is measured as the number of white pixels (constituting the object) on the CCD. Assuming a pinhole camera, the d_z and ξ_A are related as

$$d_z = f \sqrt{\frac{A}{\xi_A}}. \tag{10.4}$$

Note that f and A are constants for one particular lens and object.

Given the placement of the camera, and the data that are obtained from it, the visual processing system can be discussed in more detail. The system must accomplish the following tasks [113]:

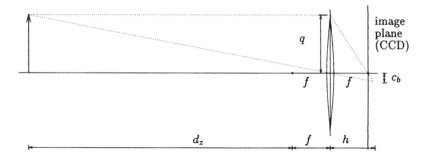

Figure 10.2. An optical system with the CCD placed at the focal distance from the lens. The symbols are referred to in the text.

(i) *Image acquisition.* The image, projected on the camera's image plane, must be transferred to the memory of the image processing hardware. This renders a discretized image $\mathcal{I}$.

(ii) *Image segmentation.* From the discrete image $\mathcal{I}$ it is determined which parts represent components, and which represent background. This is typically done using the basic principles of (dis-) continuity (edge-based) and homogeneity (region-based) techniques.

(iii) *Image description.* For the purpose of component recognition and for subsequent use in the control algorithm, for each component, features ξ describing the properties of the component as well as its position are determined.

(iv) *Component recognition.* The identified components are labelled as target object, robot hand and background. For the purpose of this chapter, only the target object and the background need to be identified.

The visual system used for the experiments described in this chapter needs only identify the three-dimensional position of an observed object; the depth is calculated from the area of the object.

10.2.3 Control systems

Having qualitative models of the sensors and the robot, a controller must be designed to solve the task set out in the introduction.

The design of a controller depends on the knowledge that is available from the process that is to be controlled. Three stages of control are distinguished [204]:

(i) When the controller has complete information about the behaviour of the inputs of the process, and this process is fully specified, the process is called a *deterministic control process.*

(ii) When unknown variables are present as random variables with known distribution functions, the process is called a *stochastic control process*.

(iii) Finally, when even that information is not available, but the controller must learn to improve its performance by observing the behaviour of the controlled process, this process is referred to as an *adaptive control process*.

Clearly, within the scope of the task set out in the introduction, the process that is controlled is in the third of the above categories.

An adaptive controller has the problem of the following duality [204]: firstly, it must *identify* the process that is to be controlled and find its parameters, and secondly, it must find which actions are required for *control* of the process. Two solutions exist [205]: *direct adaptive control* and *indirect adaptive control*. In indirect control (figure 10.3(b)), the parameters of the plant (i.e. the system that is controlled) are estimated on-line, and these estimates are used to update the parameters of the controller. In direct control, however, plant parameters are not estimated but the control parameters are directly updated to improve the behaviour of the system (figure 10.3(a)).

The direct control method in figure 10.3(a) works as follows. The plant is controlled by a signal u, and outputs a signal y_p. The *reference signal* is a signal r which is input to the controller; r can be regarded as the *desired situation* or *desired state* of the system. The *reference model* is used to compute the desired plant output y_m from the reference signal or *setpoints* r. The reference model translates r to the domain of y_p, resulting in y_m. The task for the controller C is to generate a signal $u = C(r)$ which minimizes $\|y_m - y_p\|$, i.e. it minimizes the difference between the actual and the desired situation. This error signal is subsequently used in the update of the controller.

The direct control method is normally not used for the following reason. The error signal $\|y_m - y_p\|$ carries information on how the output of the *plant* must change, and not the output of the *controller*. In order to adapt the controller, however, the error must be available in terms of u. This can only be determined when $\partial u / \partial y_p$ is known, which requires an inverse model of the plant. This is a serious drawback of this approach, since this inverse model is usually not available.

10.2.3.1 Structure of the proposed controller

However, the direct control method can still be used when no reference model is needed. This is the case when the reference signal r is expressed in quantities which can directly be measured from the plant, i.e. r and y_p are in the same domain.

The robot–camera system can be seen as a discrete state machine, where the controller C is assumed to be delay-free, and we can write $\xi_d[i + 1]$ instead of r, and $(\theta[i + 1], \xi[i + 1])$ for y_p. Since the robot moves from internal joint state to $\theta[i + 1]$, we will denote the controller output which effectuates this

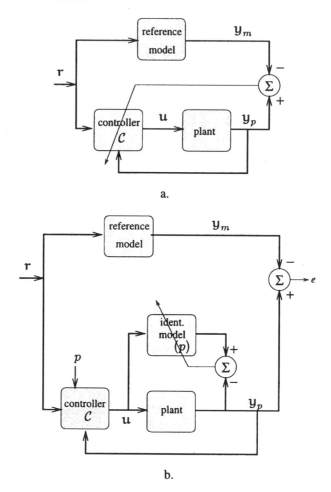

Figure 10.3. (a) Direct adaptive control. (b) Indirect adaptive control.

move by $\Delta\theta[i]$. Also, note that $\xi[i+1]$ is the signal for which $\xi_d[i+1]$ is the desired value. This means that, when $\xi[i+1] = \xi_d[i+1]$, a 'successful' $\Delta\theta[i]$ had been generated and applied to the robot. Differently put, we know that

$$C^0(\theta[i], \xi[i], \xi_d[i+1]) = \Delta\theta[i]$$

where C^0 is the ideal controller. In all other cases a valid transaction is still available, even though $\xi[i+1] \neq \xi_d[i+1]$. Thus we can focus on direct controls only, and do not require an identification model. After all, the plant output and the controller input are in the same domain. The resulting controller is depicted in figure 10.4.

The studied direct adaptive controller consists of a feedforward neural

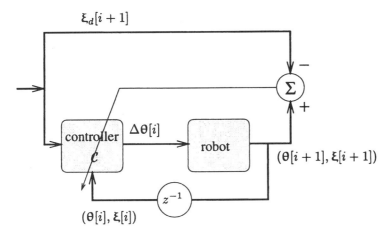

Figure 10.4. Structure of the proposed controller. After receiving input and the desired value, the controller has enough information to determine the plant input.

network trained with conjugate gradient optimization. We will refer to this neural controller by the symbol N instead of C. The universal approximation capabilities of such networks allow them to be used in a direct control scheme while eliminating the requirement for a reference model.

10.2.3.2 Generating learning samples

The task of the network is to learn the relationship between the transition from $\xi[i]$ to $\xi[i+1]$ on the one hand, and $\theta[i]$ to $\theta[i+1]$ on the other. The neural network must be trained to generate robot motion commands $\Delta\theta[i]$ to control $\xi[i]$ towards $\xi_d[i+1]$. To teach the neural network about the transition we use the *learning by interpolation* method.

In learning by interpolation [323], besides the robot state $\theta[i]$ not only the reference signals $\xi[i]$ are input to the neural controller, but also their desired values at $i+1$, $\xi_d[i+1]$. The ideal controller C^0 would generate a correct $\Delta\theta[i]$ which makes $\xi[i+1] = \xi_d[i+1]$. When the $\xi[i+1] \neq \xi_d[i+1]$, the generated $\Delta\theta[i]$ was not correct, but still it is known that the camera–robot system makes a transition from $\xi[i]$ to $\xi[i+1]$ which can be used as a learning sample. The closer $\xi[i+1]$ and $\xi_d[i+1]$, the more useful information this transition carries. In any case, a learning sample is available: input $(\theta[i], \xi[i], \xi[i+1])$ maps on $\Delta\theta[i]$.

This process of generating learning samples is illustrated in figure 10.5.

The learning samples will, in general, be situated 'around' the line where $\xi[i+1] = \xi_d[i+1]$, but by interpolating those learning samples, an approximation for that line is assumed to be generated.

Note that the dimensionality of the input space had been increased by

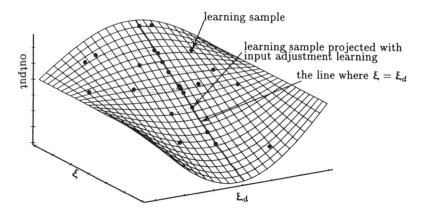

Figure 10.5. Learning by interpolation. All learning samples are positioned on a hyperplane (in the figure shown in two dimensions). The only part of the hyperplane that is used for control of the robot arm describes, in the figure, a line. Because of the increased dimensionality of the plane, the learning samples are distributed around the line, instead of on it.

the dimensionality of ξ, leading to a sparseness of the learning samples. This sparseness can be understood from figure 10.5. Samples are positioned on the hyperplane, whereas only the values situated on the 'hyperline' are used by the controller.

10.3   Using visual feedback in robot control

10.3.1   An experimental setup I

10.3.1.1   Introduction

As concluded in the previous section, the visual observation that is minimally required to grasp the object consists of the observed position (ξ_x, ξ_y) and area ξ_A of an object projected on the camera's CCD. We will write $\xi = (\xi_x, \xi_y, \xi_A)$ to denote the visual observation. Secondly, the position of the robot is described by its joint values $\theta = (\theta_1, \theta_2, \theta_3)$.

In this chapter, a neural controller will be constructed which, given the state (ξ, θ) of the camera–robot system, generates a *robot setpoint* $\Delta\theta = (\Delta\theta_1, \Delta\theta_2, \Delta\theta_3)$, to move the system *to the goal state*, i.e. reaching the state where

$$\xi = \xi_d. \tag{10.5}$$

When this state is reached, we say that the *goal* is attained.

10.3.1.1.1 Open loop control.   We are trying to construct an adaptive controller which generates a robot joint rotation $\Delta\theta$ such that, when this rotation is applied to the robot, the observed object will be located in the centre of the camera image at the predefined size. Using the symbol N for the neural controller mapping which we are looking for, and $\mathcal{R}$ for the given robot–camera mapping, the following two transformations are performed:

$$N(\xi, \theta) = \Delta\theta$$

such that

$$\mathcal{R}(\xi, \theta, \Delta\theta) = (\xi_d, \theta')$$

where $\xi_d \equiv (0, 0, A)$ and $\theta' = \theta + \Delta\theta$. When N indeed generates the correct joint rotation $\Delta\theta$ for all possible combinations of ξ and θ, we call N the *ideal* controller and write C^0. So, by definition

$$\forall \xi, \theta : \quad \mathcal{R}\left(\xi, \theta, C^0(\xi, \theta)\right) := (\xi_d, \theta').$$

The neural controller N consists of a feedforward neural network. N is being trained from learning samples $(\xi, \theta; \Delta\theta)$ from which it is known that

$$C^0(\xi, \theta) = \Delta\theta.$$

The method for creating these examples is explained below. In order to find optimal weight values in the network, the Polak–Ribière conjugate gradient algorithm with Powell restarts is used [322].

10.3.1.1.2 Closed loop control.   However, N will in general not be equal to C^0. There will always, depending on the structure of the neural network and the optimization method chosen, remain an error in the approximation to the underlying function. It is therefore likely that, when controlling the robot arm in an open loop, the hand-held camera loses track of the object when it is approached. A single joint rotation which places the end-effector only a few centimetres away from the target location will mean that the camera does not see the object any longer because it is outside the camera's field of view, such that no information at all is available about the correctness of the previous move. By using closed loop control, thus adapting the path towards the object during the move, this problem is solved. We introduce feedback in the control loop as follows. To indicate the sequence of control, the variables ξ, θ, and $\Delta\theta$ will be time-indexed. Thus we write,

$$N(\xi[i], \theta[i]) = \Delta\theta[i].$$

Now, instead of waiting for the robot to complete the move $\Delta\theta[i]$ until it is finished, *during* that move the state of the robot–camera system is measured anew and fed into the neural controller:

$$N(\xi[i+1], \theta[i+1]) = \Delta\theta[i+1].$$

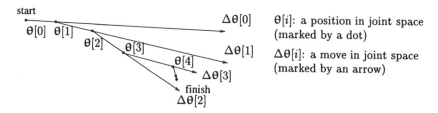

Figure 10.6. 2D motion plan of the robot arm. A planned move $\Delta\theta$ is, in general, not completed. While the robot is moving towards the new setpoint, a new setpoint is received, and the motion plan is updated.

The resulting joint rotation $\Delta\theta[i + 1]$ is sent to the robot, which immediately changes the trajectory it follows. This scheme is repeated until the target object is reached. Figure 10.6 illustrates the feedback control influence on the trajectory followed by the robot arm.

Note that the delay between iteration i and $i+1$ is not defined; it is dictated by the visual processing and communication delays.

Why will the system be more accurate when a feedback loop is introduced? There are two reasons:

(i)   Sensor readings close to the target are more accurate.
(ii) The approximation to C^0 close to the target can be made more accurate. This is done by increasing the sample density close to the goal state. This is automatically obtained, since a successful grasping trail always ends in this goal state, thus creating samples in that part of the input space.

The proposed experimental setup is shown in figure 10.7.

At time i, the visual and robot state are fed into the neural network, which generates a joint rotation. That rotation is realized by the inverse robot dynamics module, which calculates torques to make the robot move. One time slot later, new visual and robot state data are available and given to the neural network.

10.3.1.2   Constructing the controller

In this section, the details of the implementation are discussed. A description of the input/output behaviour, as well as a typical control loop, is given.

10.3.1.2.1   The neural network.
The neural network consists of a feedforward network trained with conjugate gradient backpropagation as described in [322]. The visual inputs ξ and robot state inputs θ together make five network inputs (as discussed before, the structure of the robot eliminates the use of θ_1 as network input). Three outputs $\Delta\theta_1$, $\Delta\theta_2$, and $\Delta\theta_3$ constitute the network output and are given to the robot as a rotation (delta joint value) for joints 1, 2, and 3. The visual input consists of the position (ξ_x, ξ_y) measured in pixels relative to the

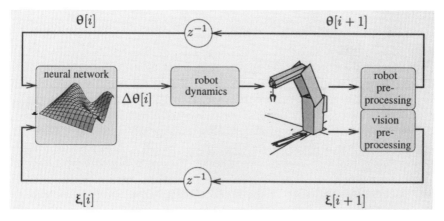

Figure 10.7. The experimental setup. The visual state $\xi[i]$ and robot state $\theta[i]$ are input to the neural controller, which generates a joint rotation $\Delta\theta[i]$. The robot's PID controller (marked 'robot dynamics' in the figure) calculates the required torques to make the robot move. At the next step $i + 1$, the new robot and visual state are available.

camera centre, plus the area ξ_A of the object, also measured in pixels. With specialized hardware these quantities are measured in real time, i.e. within the time needed for grabbing one frame. At the time a frame is grabbed, and before the position and area of the object are extracted, the robot position is measured and stored in a register.

The neural network must learn the relationship between displacements in the visual domain and displacements in the robot domain. This relationship is contained in the measured visual data $\xi[i]$ and $\xi[i + 1]$ in relation to the robot data $\theta[i]$ and $\theta[i + 1]$.

Knowing that a robot move from $\theta[i]$ to $\theta[i+1]$ corresponds to an observed object move from $\xi[i]$ to $\xi[i + 1]$, these data can be used in the construction of learning samples which describe the actual behaviour of the camera–robot system.

The conjugate gradient method that is used to update the weights in the neural network minimizes the summed squared error over a *set* of learning samples. Therefore the learning samples which are generated during control of the camera–robot system are collected in *bins*.

10.3.1.3 *Using bins*

When n steps are used to move towards the object, $n \times (n-1)$ learning samples can be constructed. This can be seen from figure 10.6: for instance, apart from data obtained from the move $\theta[0] \rightarrow \theta[1]$ and $\theta[1] \rightarrow \theta[2]$, also the move $\theta[0] \rightarrow \theta[2]$ can be constructed by combining the previous two moves. With a typical value of n set to 100, a single trial may lead to nearly 10 000 samples.

Clearly, the number of samples would grow quickly out of bounds, leading to unacceptable delays in the learning algorithm, when all learning samples were kept.

Therefore a selective binning structure is set up as follows. Along each input dimension d of the neural network ($1 \leq d \leq 5$ in this case) a partition into $b[d]$ parts is made. This partitioning leads to a five-dimensional structure of bins: a hypercube. When a learning sample is available, its input values uniquely determine which bin in the hypercube this learning sample fits in. When the corresponding bin is empty, the new sample is put in it; otherwise the sample in that bin is replaced by the new sample. Thus each bin contains only one learning sample.

An obvious advantage of the hypercube method to store the samples is that a uniform or otherwise desired distribution of the learning samples can be realized. This means that the neural network approximation will not lose accuracy in those parts of the input space which have not been visited for a long time.

10.3.1.4 *Learning samples*

The neural network N is trained with learning samples $(\xi, \theta; \Delta\theta)$. Unfortunately, it is not possible to analytically compute a $\Delta\theta$ from a given (ξ, θ), since that would require knowledge of the ideal controller C^0. How can learning samples be constructed? In learning by interpolation the inputs to the neural controller do not only consist of $\theta[i]$ and $\xi[i]$ but also ξ_d. The network, which we will indicate by N', thus has eight inputs and three outputs. The task of the network is to generate a joint rotation

$$N'\left(\xi[i], \theta[i], \xi_d\right) = \Delta\theta[i]$$

such that

$$\mathcal{R}\left(\xi[i], \theta[i], \Delta\theta[i]\right) = \left(\xi_d, \theta[i+1]\right).$$

The learning samples that are gathered, however, give information on how the robot moves from $\xi[i]$ to $\xi[i+1]$, where $\xi[i+1]$ need not coincide with ξ_d. Thus the neural network learns the mapping

$$N'\left(\xi[i], \theta[i], \xi[i+1]\right) = \Delta\theta[i]$$

such that

$$\mathcal{R}\left(\xi[i], \theta[i], \Delta\theta[i]\right) = \left(\xi[i+1], \theta[i+1]\right).$$

This has the advantage that the neural controller can be used to move the robot to *any* target point, and that certain systematic errors (e.g. from the camera) are taken care of.

10.3.1.5   *The control loop*

In the first setup we are faced with the following system.

(i)   Set $i = 1$, and set all measurements at $i = 0$ to zero. Set $\Delta\theta[1]$ to a small random value.

(ii) The robot control command (desired joint position) $\theta_d[i]$ is sent to the robot, and the robot moves.

(iii) The robot records the current joint values $\theta[i]$, as well as an image $\mathcal{I}$ from which the visual data $\xi[i]$ are extracted. The ξ consists of $\xi = (\xi_x, \xi_y, \xi_A)$ where
  $\xi_x = x$ coordinate of target object in image
  $\xi_y = y$ coordinate of target object in image
  $\xi_A = $ observed area of target object in image.

(iv) We know that the robot moved from joint position $\theta[i - 1]$ with joint rotation $\Delta\theta[i]$, and at the same time the visual data changed from $\xi[i - 1]$ to $\xi[i]$. From this we can construct the following learning sample:

$$C^0\big(\xi[i - 1], \xi_d[i], \theta[i - 1]\big) = \Delta\theta[i].$$

(v)   The neural network is fed the visual and positional information $\xi[i]$ and $\theta[i]$, as well as the desired visual data at $i + 1$, $\xi_d[i + 1]$. In this case, θ is a *context variable* while ξ is the *control variable*. The network generates an output

$$N'\big(\xi[i], \xi_d[i + 1], \theta[i]\big) = \theta_d[i + 1].$$

(vi) $i \leftarrow i + 1$; goto (ii).

10.3.1.6   *Results*

Results are shown in figure 10.8. The average error in the first step is less than 1 cm and in two steps approximately 1 mm. Note that after a few iterations, the target can already be reached in the feedback loop; such fast learning makes the method very well suited for real applications.

Secondly, the method's adaptability is tested. The state of the network after the initial 800 goals is taken as the beginning state. The system is changed by rotating the hand, and hence the hand-held camera, by 45°. The result is shown in the bottom row of figure 10.8.

10.3.2   An experimental setup II

10.3.2.1   *Introduction*

The approach shown in the previous section has been demonstrated to be successful in practice. However, we made one simplification: it was assumed that the dimensions of the target object were known.

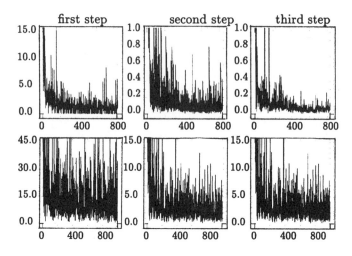

Figure 10.8. Grasping error in cm with the simulated robot after 1 (left column), 2 (second column), and 3 (third column) steps. On the horizontal axes, the number of goals is plotted. The desired accuracy was a Euclidean distance of 0.5 cm between the end-effector and the target object. Top row. Initial learning: results when learning from scratch. Bottom row. After initial learning: rotation of the camera by 45°.

When the dimensions of the object are unknown, and vary from one run to another, this approach cannot be used. However, we can still use monocular vision, by the use of optic flow.

10.3.2.2 Optic flow

Optic flow, which is defined as the motion of the observer's environment projected on its image plane, is in fact more commonly used by living organisms than information obtained from static retinal patterns. The fact that optic flow is fundamental to vision has only been realized since the pioneering work of Gibson [121]. For instance, the gannet (figure 10.9), when feeding, dives down in the water. During the dive, the bird needs its wings to adapt its path to the motion of the fish; however, at the moment of contact with the seawater its wings must be folded to prevent them from breaking. It has been shown that the time remaining between the moment that the bird folds its wings, and when it hits the water, is always the same for one particular bird. It is not controlled by height or velocity separately, but by the quotient of the two. This remaining time is known as the *time-to-contact* and indicated by the symbol τ. When the system is not accelerating, τ is given by the quotient of the distance from an approaching surface and the velocity towards it. Exactly the same information can be obtained from the divergence of the approaching surface [171], a feature that can be observed monocularly.

Figure 10.9. The gannet.

Since this bird cannot measure its velocity, it measures the time until it hits the water from time derivatives of the visual observation. It is this mechanism that underlies the method presented in this chapter for controlling a monocular robot arm such that it 'lands' on an object. An ordinary grasping task can be described as: 'at some time, the distance between the object and the hand-held camera must be zero'. We go one step beyond that requirement: the distance must be zero *at some time* (which, we now know, is related to the time-to-contact), and while the system is at rest, the velocity, acceleration, and higher derivatives must also be zero. We will call this the *goal state*. But this can be seen as the endpoint of a trajectory towards that point in time. In the case of the bird, the decision to fold its wingspan can be made with the available visually measured quantities. Below it will be shown that by extending the above example to higher-order time derivatives of the visual data, criteria can be developed which specify a trajectory which ends in a rest state (i.e. zero velocity, acceleration, etc) at the end point. These criteria will lead to *visual setpoints* along the followed trajectory, and are used as inputs to a neural controller which must generate robot joint accelerations in order to follow the setpoints in a visual domain. Thus it is possible that the eye-in-hand robot arm exactly stops on an observed object by using optic flow. By using time derivatives of the visual field we obtain an important advantage: the model of the object is not needed to obtain depth information. The system need not be calibrated for each object that has to be grasped, but can approach objects with unknown dimensions.

10.3.2.3 Theoretical background

We can describe the motion of the camera, and with it, the robot's end-effector, as a trajectory x_r in world space. This trajectory is known in terms of the robot variables θ; i.e. we know this trajectory in terms of a sequence of robot joint positions.

Secondly, we can describe the position of the target object as x_o; this position is unknown and cannot be measured, since we have only a single camera and do not know the dimensions of the object. Recalling equation (10.4), we cannot even measure the position of the object *with respect to the camera*: if the object appears four times as small as before, it can be twice as far away, or four times as small, or any mixture of those two.

Let us define $d(t) \equiv x_r(t) - x_o$; the unknown quantity $d(t)$ is the *distance between the robot and object at time t*. The grasping task can be formulated as reaching the robot position $\theta[\tau]$ at some desired time τ where $d(t \geq \tau) = 0$. That is, *at some time τ and thereafter the position must be zero.* This can be realized when we require that $d(\tau)$ as well as its derivatives are zero at τ; after all, when its derivatives are zero at that moment, $d(\tau)$ will remain zero. The system will remain in rest. We call this the *stopping criterion*.

Assume that we describe each of the components of $d(t)$ (i.e. the x, y, and z component as well as any rotational components) by a Taylor series,

$$d(t) = \sum_{j=0}^{n} a_j t^j + \epsilon. \tag{10.6}$$

The above stopping criterion can now be expressed as

$$\forall 0 \leq k < n : d^{(k)}(\tau) = \sum_{j=k}^{n} \frac{j!}{(j-k)!} a_j \tau^{j-k} = 0. \tag{10.7}$$

It can be shown [324] that this leads to the following constraints on the parameters a_j:

$$\forall 0 \leq k < n : \frac{a_n^{n-k-1} a_k}{a_{n-1}^{n-k}} = \frac{1}{n^{n-k}} \binom{n}{k}. \tag{10.8}$$

10.3.2.3.1 Getting the visual data.   Now, how are the a_j related to the visual data that we can measure? Taking (10.4) into account again, we know that we can visually determine

$$\xi_x(t) = f \frac{d_x(t)}{d_z(t)} \qquad \xi_y(t) = f \frac{d_y(t)}{d_z(t)} \qquad \xi_A(t) = \frac{f^2 A}{d_z(t)^2}.$$

When we define

$$\xi_x'(t) \equiv \frac{d_x(t)}{\sqrt{A}} \qquad \xi_y'(t) \equiv \frac{d_y(t)}{\sqrt{A}} \qquad \xi_z'(t) \equiv \frac{d_z(t)}{f\sqrt{A}} \tag{10.9}$$

then again we can make Taylor polynomials for $\xi(t)$ such that

$$\xi(t) = \sum_{i=0}^{n} v_i t^i + o(t^n) \tag{10.10}$$

where $\xi(t)$ is either $\xi_x'(t)$, $\xi_y'(t)$, or $\xi_z'(t)$. Similarly, v_i indicates the x, y, or z parameters.

Once the parameters v_i are known, the polynomials for $\xi'_x(t)$, $\xi'_y(t)$, and $\xi'_z(t)$ are known. Knowledge of these parameters, however, does not give sufficient information on the position, velocity, etc of the end-effector relative to the object, since they are scaled by the constants A and f. However, the constraints can still be expressed in visual parameters: using the polynomial expansions for $d(t)$ and $\xi(t)$, and combining these with equations (10.9), the v_i have a common constant

$$v_i = c a_i \tag{10.11}$$

where c is c_x, c_y, or c_z for the x, y, and z components of d, given by

$$c_x = A^{-1/2} \qquad c_y = A^{-1/2} \qquad c_z = (f^2 A)^{-1/2}. \tag{10.12}$$

When we take (10.8) into account, we see that it can be rewritten as

$$\forall 0 \leq k < n : \frac{v_n{}^{n-k-1} v_k}{v_{n-1}{}^{n-k}} = \frac{1}{n^{n-k}} \binom{n}{k}. \tag{10.13}$$

Since the equation contains the fraction c^{n-1}/c^{n-1}, the unknown constants disappear and we can determine (10.13) from visual measurements.

However, there are some problems with this equation. Firstly, in a real-world system we would usually try to control second- or third-order trajectories; i.e. $n = 2$ or $n = 3$. This means that we have to visually determine v_0, v_1, v_2, and perhaps v_3. Knowing that the visual frame rate is low, and that visual data contain large amounts of noise, calculating v_2 is tough and v_3 is prohibitive.

Secondly, we cannot control the time that the trajectory lasts. This is determined by the initial conditions: for instance, if the robot initially moves fast and is not far away from the object, the trajectory will be quickly traversed. This poses a problem when the trajectory time differs for the x, y, and z components of the motion; it would be better if the robot arm moved in a straight path towards the object.

To solve both of these problems we introduce an extra constraint. Since $d(t)$ is approximated by a polynomial of order n, the nth derivative of the approximation of $d(t)$ must be constant in time; a_n is constant. Therefore, the $(n - 1)$th must be linear in time. Consequently, the time to bring the $(n - 1)$th derivative to 0 is equal to the quotient of the two. For $n = 2$, this is the velocity divided by the acceleration. In the general case,

$$\tau = -\frac{a_{n-1}}{n a_n}. \tag{10.14}$$

This can be combined with the constraints (10.8), such that the system is now faced with n non-trivial constraints:

$$\frac{a_n{}^{n-k-1} a_k}{a_{n-1}{}^{n-k}} = \frac{1}{n^{n-k}} \binom{n}{k} \qquad 0 \leq k < n \qquad \text{and} \qquad \tau_d = -\frac{a_{n-1}}{n a_n}. \tag{10.15}$$

These constraints can be rewritten as

$$\frac{a_k}{a_{n-1}} = (-\tau_d)^{n-k-1} \frac{1}{n} \binom{n}{k} \qquad 0 \le k \le n. \tag{10.16}$$

Similar to the time-independent case, satisfying these n non-trivial constraints leads to the desired trajectory. However, the constraints are all related and a simplification is in order.

Consider once again the polynomial expansion of $d(t)$ in (10.6). This polynomial expansion of order n is globally valid over the whole time interval, i.e. the whole trajectory of the robot arm. After splitting up the global time axis in intervals, the $d(t)$ can be *repeatedly* approximated in these intervals by polynomials with parameters $a_j[i]$. These approximations are written as

$$d[i](t[i]) = \sum_{j=0}^{n} a_j[i]t[i]^j + \epsilon. \tag{10.17}$$

Note that $d(t) \equiv d[0](t[0])$, but that the parameters $a_j[i]$ are in general not equal to $a_j[i+1]$! The starting time t at which the $d[i]$ and thus the $a_j[i]$s are defined is repeatedly changed.

As set out above, the task of the feedforward based neural controller is to make the robot manipulator follow a pre-specified trajectory. During a time interval i the system measures the $\xi[i]$ (note that, due to the discrete-step feedback loop, the ξ are now discrete variables indexed by the step number instead of varying continuously in time). From these measurements the controller generates joint accelerations $\ddot{\theta}[i+1]$ and sends them to the robot. This marks the beginning of a new time interval $i+1$.

Now, by using the repeatedly updated $d[i]$, we can combine the constraints (10.15) and find the simplified form

$$\frac{a_0[i]}{a_1[i]} = -\frac{(\tau_d - t[i])}{n} \qquad \forall i : 0 \le i < v \tag{10.18}$$

where $v \ge n$ and $(\tau_d - t[v]) \ge 0$. The proof of this theorem can be found in [324]. We will refer to (10.18) as the *time-dependent constraint*.

The time-dependent constraint is obtained by extending the *local* time intervals towards the moment when $(\tau_d - t[i]) = 0$. Although the $d[i](t[i])$ is a local approximation, we can just pretend that it fits the *whole* $d(t)$ starting at $t[i] = 0$, and let the time-dependent constraint be valid for that trajectory.

τ_d is usually chosen equal for the x, y, and z components of $d(t[i])$, to ensure that all components go to zero at the same time.

10.3.2.4 The control loop

The time-dependent constraint (10.18) is used in a controller as follows.

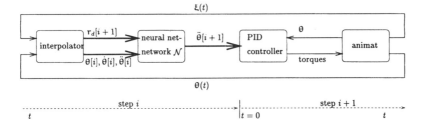

Figure 10.10. The structure of the time-to-contact control loop.

(i) Set $i = 1$, and set all measurements at $i = 0$ to zero. Set $\ddot{\theta}_d[1]$ to a small random value.

(ii) The robot control command (desired joint acceleration) $\ddot{\theta}_d[i]$ is sent to the robot, and the robot moves.

(iii) The robot records the current joint values $\theta[i]$ from which the $\dot{\theta}[i]$ and $\ddot{\theta}[i]$ are computed, as well as an image $\mathcal{I}$ from which the visual data $\xi[i]$ are extracted. The ξ consists of $\xi = (\xi_x, \dot{\xi}_x, \xi_y, \dot{\xi}_y, \xi_A, \dot{\xi}_A)$ where
$\xi_x = x$ coordinate of target object in image
$\xi_y = y$ coordinate of target object in image
$\xi_A =$ observed area of target object in image.

(iv) Set $r[i] = a_0[i]/a_1[i] - (\tau_d - t[i])/n$ for the x, y, and z directions.
We know that the robot moved from joint position $\theta[i-1]$ with joint rotation $\theta_d[i]$, and at the same time the visual data changed from $\xi[i-1]$ to $\xi[i]$. From this we can construct the following learning sample:

$$C^0\big(r[i-1], r[i], \theta[i-1], \dot{\theta}[i-1], \ddot{\theta}[i-1]\big) = \ddot{\theta}_d[i].$$

(v) Apply the newly measured data to the network. The network generates an output

$$N\big(r[i], 0, \theta[i], \dot{\theta}[i], \ddot{\theta}[i]\big) = \ddot{\theta}_d[i+1].$$

(vi) $i \leftarrow i + 1$; goto (ii).

The control loop is depicted in figure 10.10.

10.3.2.5 Results

In order to measure the success of the method while applied to the simulated robot, we measure the $d(t)$, $\dot{d}(t)$, and $\ddot{d}(t)$ during the trajectory; with the simulator, these data are available. A correct deceleration leads to a $d(t) = \dot{d}(t) = 0$ when $\tau_d = 0$, i.e. at the end of the trajectory. The results of a run with the simulated OSCAR robot are shown in figure 10.11. This graph shows the distance between the end-effector and the target object at $\tau_d = 0$.

The results show that after only a few trials (in this case, 4), the positional error at $(\tau_d - t[0]) = 0$ is below 1 mm, while the velocity is below 0.1 cm

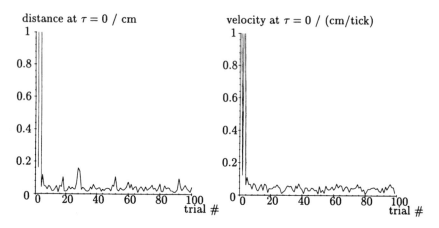

distance at $\tau = 0$ / cm

velocity at $\tau = 0$ / (cm/tick)

trial #

trial #

Figure 10.11. Distance and velocity at $\tau = 0$. The left figure shows the distance $\sqrt{d(\tau = 0)_x^2 + d(\tau = 0)_y^2 + d(\tau = 0)_z^2}$ between the end-effector and the approached object. The right figure shows the velocity $\sqrt{\dot{d}(\tau = 0)_x^2 + \dot{d}(\tau = 0)_y^2 + \dot{d}(\tau = 0)_z^2}$ of the end-effector in cm/tick. Typical end-effector velocities during the trajectory are between 0.5 and 2.0. The horizontal axis indicates the trial number.

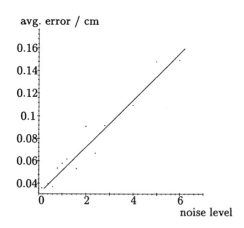

avg. error / cm

noise level

Figure 10.12. Results of the simulation when noise in the visual input is applied. The figure shows the distance $\sqrt{d(\tau = 0)_x^2 + d(\tau = 0)_y^2 + d(\tau = 0)_z^2}$ between the end-effector and the approached object averaged over 200 goals (marked by dots in the graph). The vertical axis shows the noise level l. The figure clearly shows that the noise level and the grasping error are linearly related. At values of $l \geq 7$ the signal-to-noise ratio is so large that the system cannot always reach the target at $\tau_d = 0$, but sometimes overshoots. The average error at $l = 7$ goes up to 2.0 cm; at $l = 8$ it is as high as 4.0 due to overshoots.

per simulator time unit (typical end-effector velocities during deceleration are 0.5–2.0 cm per simulator time unit).

Figure 10.12 shows the results of the control algorithm when tested with different noise levels. Noise is present in the measurements, and also in the learning samples which are taught to the neural network. A graceful degradation is shown up to very high noise levels, as high as seven times the expected amount of noise.

Acknowledgments

Continuing support by Gerd Hirzinger, who leads the Department of Robotic Systems at the German Aerospace Research Establishment, is kindly acknowledged. The author is grateful for the many fruitful discussions with Frans Groen which were essential for this work.

Chapter 11

Designing a Nervous System for an Adaptive Mobile Robot

Tom W Scutt [1,3] and Robert I Damper [2]

[1] University of Nottingham, UK

[2] University of Southampton, UK

[3] tom.scutt@nottingham.ac.uk

11.1 Introduction

Historically, the original motivation for building robots was to facilitate industrial automation (see [114]). Indeed, the word *robot* derives from the Czech *robota* meaning 'work'. In recent years, however, increasing attention has been paid to the possibilities which robots—especially mobile, learning robots—offer for modelling intelligent behaviour. Thus, designing and building robots in order to observe their actions in the real world has become an important tool of the cognitive scientist. The publication of Braitenberg's small but influential book *Vehicles* [45] in 1984 arguably marks the dawn of this new era of *synthetic psychology* (although his experiments were of the *gedanken* variety—no physical systems were built), with Brooks (e.g. [50]) probably its acknowledged current flag-bearer.

While it is traditional to think of robots as embodying sensori-motor control systems, there is a very real sense in which the robot vehicles which Braitenberg described possessed a rudimentary *nervous system*. In the simpler cases, a vehicle might be little more than two spatially separated light sensors whose outputs directly control the speed of two motors, one on each side, driving the vehicle. Hence, the nervous system merely consists of two sensory and two motor neurons, with direct connections between them. By exploiting the small range of possibilities for excitatory/inhibitory, homolateral or cross-coupled connections, a variety of light-seeking or light-avoiding behaviours emerges—and Braitenberg is greatly preoccupied with a hypothetical observer's assessment

220

of these behaviours in intentional terms. In other cases, the nervous system involves non-linear dependences, activation functions with discontinuities and thresholds, or networks of McCulloch–Pitts [194] synthetic neurons, leading to ever more complex behaviours. Indeed, Braitenberg himself points out ([45], p 20) that observed behaviour often 'goes beyond what we had originally planned', while Dewdney [88] writes: 'Braitenberg's vehicles teach us that even the most primitive nervous systems [*are*] capable of behavior that seems complicated or surprising'. Again as pointed out by Braitenberg: 'It is actually impossible in theory to determine exactly what the hidden mechanism is without opening the box [*that contains the vehicle's nervous system*], since there are always many different mechanisms with identical behaviour'. This 'theory' will be familiar to physicists and engineers as the Helmholtz–Thévenin principle, according to which one cannot uniquely determine the contents of a 'black box' physical system from its external ('terminal') behaviour. The principle, which was expounded by Helmholtz in the context of electro-magnetic fields and by Thévenin in the context of electrical circuits, deserves to be more familiar to psychologists!

If our goal is to understand human (or animal) behaviour, this leads us to a major problem. Analysis is not a powerful enough tool to cope with the behaviour of even the simplest of 'black box' artificial systems, let alone the complexities of the human mind. This is the attraction of synthesis—*à la Vehicles*—since, in this case, we know in precise detail the physical substrate for the observed behaviour. Yet while it is 'pleasurable and easy to create little machines that do certain tricks' ([45], p 20) these creations need to bear some systematic relation to the realities of biology and psychology if they are to tell us anything about natural, intelligent organisms.

In this chapter, we develop the hypothesis that, by studying neurophysiology (and, in particular, small systems of neurons [155] in lower animals), we can discover a *sensible* starting point for the pursuit of synthetic psychology— i.e. the synthesis of behaviour in adaptive mobile robots that simulates behaviour in natural systems in some useful and meaningful way. The philosophical underpining of our work owes much to Barlow's famous 'neuron doctrine' [35]. Our approach is unashamedly holistic, cf. Lloyd's 'simple minds' [179]. The complexity of behaviour exhibited by Braitenberg's vehicles arises because they are complete input/output systems grounded in the real world. If we were to cut the link between sensory and motor units, there would be little or nothing left to study. This is counter to the reductionist, 'modularity of mind' orthodoxy (see Fodor [107]) of a few years ago, which held that intelligent behaviour results from the interaction of mental phenomena which can be—and indeed must be—experimentally isolated and studied individually. While we believe (with, e.g. Franklin [110]) that the bottom-up and top-down approaches are complementary and should be developed together, Fodor is clearly convinced that his is the only right way. For instance, in 'The folly of simulation' [108], he states: ' … any attempt to build intelligent machines is … an engineering problem and has no intrinsic scientific interest... I don't see the scientific interest of simulating behavior'. A major purpose of our work is to demonstrate the scientific insights which can be gained by implementing simple robots situated in the real world in order to study their behaviour (see also [69]).

The remainder of this chapter is structured as follows. In section 11.2, we contrast two of the possible approaches to the design of a nervous system for an adaptive robot—currently popular 'parallel distributed processing' (PDP) and biologically motivated neural modelling—concluding that the latter holds more promise, and that adaptive behaviour should be based on neurophysiology of learning, as understood from the study of small systems of neurons in lower animals. Section 11.3 briefly outlines the essentials of this latter topic. In section 11.4, we describe the neurophysiology of so-called central pattern generators—a type of neuron responsible for rhythmic behaviour underlying, for instance, locomotion in lower animals. This illustrates the importance to behaviour of specialized neuron types—a biological fact mostly ignored in the PDP paradigm. Our approach can be characterized as fitting within the 'computational neuroscience' paradigm which is next described in section 11.5. In section 11.6, we describe the *Hi-NOON* neural simulator which was written to facilitate our work, while section 11.7 describes simulations of biologically motivated forms of learning undertaken using *Hi-NOON*. Section 11.8 details *Hi-NOON* simulations—one quite simple and the other more complex—of robot vehicles which learn from their environment. We conclude in section 11.9 with discussion of the implications of this work for modelling adaptive behaviour in general.

11.2   Modelling a nervous system

In recent years, parallel distributed processing (PDP) or 'connectionism' [344] has become the majority paradigm in cognitive modelling and (non-symbolic) artificial intelligence (AI). It greatest successes obviously lie with back-propagation: however, we believe that conventional feedforward artificial neural networks are of limited use for developing adaptive mobile robots for three reasons:

(i)   Such networks almost invariably consist of a homogeneous population of McCulloch–Pitts type neurons. Real systems of neurons are not homogeneous, but have different neuron types specialized to particular functions. In our view, it is important that our mobile robot's nervous system has heterogeneous, specialized neurons also.

(ii) Most artificial neural nets still rely on some form of back-propagation learning and, therefore, on a training signal. Hertz, Krogh and Palmer ([140], p 119), citing Crick [96], describe back-propagation as 'rather far-fetched as a biological learning mechanism'. Autonomous agents should not need to be pre-trained, at least in supervised fashion. They should discover the world and adapt to it by themselves.

(iii) Learning in conventional PDP models is typically passive and open-loop.

Elaborating on this third point, input units are activated, activation is fed forward, and after the effects of one or more hidden layers, the output units display an activation which will be dependent on the connectivity of the network. This output will then be used to calculate delta terms (if the net is in learning mode), or used by another system (typically a human observer). The output of the system does not affect the system's world. Typically, output from a network has no effect on input to the network, especially in learning mode. Compare this with a simple invertebrate nervous system or even the 'nervous system' of the simple Braitenberg vehicles mentioned earlier. In these cases, the output of the network makes a difference to the world, and changes the input that the network receives. Learning typically involves associating two inputs rather than associating an input with an output, and the input training set (if it can be called that) is determined by both the environment and the motor output of the network.

One further problem of the connectionist paradigm is its insistence on the spike rate of a neuron (usually equivalent to the activation level of a unit according to the PDP literature) as the be-all-and-end-all of neuronal information transfer. (Many PDP researchers seem to be unaware that neurons generate spikes at all.) Electrical synapses, hormonal neuromodulation, and the effects of phasing in auditory systems [260] show that there is far more to a real neural network than spike rate and weighted connections—individual spikes can be important.

All this is not to say that neural network research in the PDP style is unimportant or misguided. On the contrary, it has illuminated the field of AI (and cognitive science in general) in a way the classical paradigm never managed. But connectionism has strayed a long way from its neurophysiological roots, and fits uncomfortably with the new movement towards situated systems. Feedforward networks, despite avoiding many of the representational pitfalls of classical, symbolic AI by the use of distributed representations, hidden units responsive to microfeatures, and so on [70], still suffer from producing cognitive models which are not attached 'at both ends' to the environment. By this, we refer to the fact that most PDP modules are used for performing perceptual tasks, with the semantic interpretation of the output units left to a human observer, i.e. they are ungrounded [129].

Even the simplest of animals can adapt to and survive in its environment. Studies of small systems of neurons have been vital in understanding the relationship between neurophysiology and behaviour. What these systems have in common is that their simplicity allows the observer to see how each individual neuron in a circuit contributes to behaviour. It must be borne in mind, however, that the behaviours so far examined in this way form only a tiny fragment of an extensive and complex repertoire. The gill-withdrawal reflex in the marine snail

Aplysia, for example, has been extensively studied but this animal is involved in locomotion, finding food, eating, escaping from predators (including inking), learning (as we shall see below), and mating, to name a few of its activities.

Of most interest to the constructor of nervous systems for mobile robots are the basic neuronal mechanisms underlying *adaptation* and *internal rhythms*. It will be useful here to remind ourselves briefly of the neurophysiology involved in these mechanisms.

11.3   Neurophysiology of learning

The gill-withdrawal reflex in *Aplysia* is the classic example of neurobiology being able to explain an adaptive behaviour. This is a simple defensive reflex action in which the gill is withdrawn after a stimulus. The gill lies under the protective mantle shelf, in the mantle cavity, which terminates with a fleshy spout—the siphon. When a moderately intense stimulus is applied to the siphon, the gill contracts and withdraws into the mantle cavity. (This reflex is analogous to the response whereby a human being jerks a hand away after touching a very hot object.) The neural circuitry underlying this response has been elaborated, and a neuron-level explanation of behaviour advanced [155, 131, 132, 130]. But how does this animal, and others, learn to behave in this way?

11.3.1   Non-associative learning

Thompson [313] defines *non-associative* (or *simple*) learning as that which 'results from experience with a single type of event'. *Habituation* and *sensitization* are the most basic forms of non-associative learning.

Habituation is, perhaps, the simplest form of 'learning' possible. It occurs when an animal learns to ignore a weak repetitive stimulus which is neither rewarding or noxious. Only the intensity of a response is changed; the nature of the response itself remains the same despite its attenuation. This could be viewed as a reason why habituation should not be considered real learning. However, to the extent that the behaviour is adaptive, it is valid to include it in the class of learned responses, albeit as an elementary form. In neuronal terms, this means that if neuron A synapses onto neuron B, and A fires repeatedly, then the synaptic strength will decrease and the response of B will lessen accordingly (see figure 11.1).

Habituation may mean that a neuron which was previously firing will now fail to reach threshold. Experiments have shown that the attenuation of a response to a sensory stimulus is paralleled by a decrease of activity in motor neurons (e.g. [155]). This has been found to result from a corresponding decrease in the amount of transmitter released by the synapse. As Hawkins and Kandel [133] note:

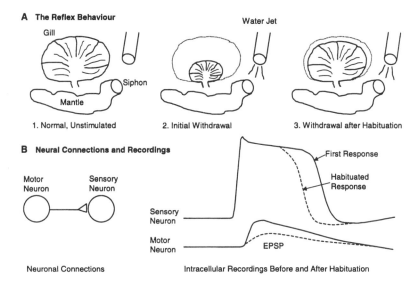

Figure 11.1. Habituation of the gill-withdrawal reflex in *Aplysia*. (A) Observed behaviour. (B) Diagram of neuronal connections and intracellular recordings showing shortening of sensory action potential and decrement in motor neuron excitatory postsynaptic potential (EPSP) with repeated stimulation (adapted from [286]).

> This depression involves, at least in part, a decrease in the amount of Ca++ [*calcium ions*] that flows into the terminals of the sensory neurons with each action potential. Because Ca++ influx determines how much transmitter is released, a decrease in Ca++ influx results in decreased release.

In *Aplysia*, this depression of the synaptic strength can last from minutes to weeks depending on the way the initial stimuli are presented.

Sensitization is a slightly more complex form of non-associative learning, in which an animal learns to respond more vigorously to a variety of previously weak stimuli after it has received a noxious or potentially dangerous stimulus. For example, if an animal is given an electric shock it will afterwards show a much greater reaction to a non-harmful tactile stimulus. Short-term sensitization lasts from minutes to hours, and can even cause dishabituation (i.e. previously habituated responses can be reinstated or enhanced). Although it might appear at first to be similar to habituation (albeit opposite in the polarity of effect), sensitization differs in several respects. It occurs in response to a stimulus different from that eliciting the initial reflex, and it involves activation of general arousal systems [286]. The neurophysiology of this effect is also rather more complicated than that responsible for habituation. Activating certain sensory neurons will, in turn, activate one or more facilitatory interneurons, which

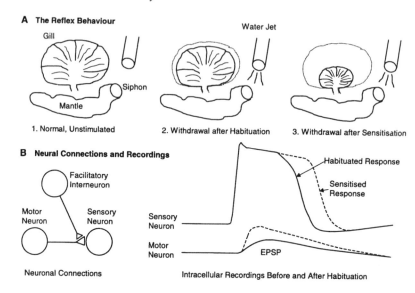

Figure 11.2. Sensitization of the gill-withdrawal reflex in *Aplysia*. (A) Observed behaviour. (B) Diagram of neuronal connections and intracellular recordings showing broadening of sensory action potential and facilitation of the motor neuron EPSP during sensitization (adapted from [286]).

themselves synapse on or near the axon terminals of other sensory neurons, not onto the neuron body itself (see figure 11.2). This is called *presynaptic facilitation*.

The transmitter released by these facilitatory interneurons decreases the number of K^+ (potassium ion) channels that are open during the action potential. This in turn leads to a broadening of subsequent action potentials, allowing a greater amount of Ca^{++} to flow into the terminal and enhancing transmitter release. The restoration of previously habituated responses results from this increase in the quantity of transmitter present in the synapse.

11.3.2 Associative learning

In *associative* learning, an animal learns to make a connection (exhibited by its behavioural response) between a neutral stimulus and a second stimulus that is either rewarding (reinforcing) or noxious (aversive). The particular form of associative learning that is discussed here is classical (or Pavlovian) conditioning. However, there are many other forms of associative learning, such as operant (or instrumental) and one-trial conditioning, although the biochemical underpinnings of these processes are far less understood than those underlying classical conditioning. (For a more detailed account of classical conditioning, see [271].)

Organisms have a number of *a priori* reflexes built into the nervous system: these are *unconditioned* reflexes. An unconditioned reflex consists of an unconditioned stimulus (US), either a reward or a punishment which triggers the reflex, and the reflex response itself, known as the unconditioned response (UR). This relation between the US and UR is demonstrated by the reaction of the dogs in Pavlov's experiment where:

$$US \text{ (food)} \rightarrow UR \text{ (salivation)}.$$

Classical conditioning occurs where another previously neutral stimulus, the conditioned stimulus (CS), produces the reflex response. The response produced by the CS is called the conditioned response (CR). This conditioning is brought about by repeatedly exposing the animal to the CS a short period before the US. The close temporal association leads to the CS eventually evoking the same response as the US.

At a neural level, classical conditioning takes place when there is a synapse-on-synapse connection, and the postsynaptic synapse fires a short period before the presynaptic one (see figure 11.3). In other words, the CS takes place shortly before the US. This leads to a large and long-lasting increase in weight of the postsynaptic synapse. This enhancement seems to reach a maximum when there is a period of about 0.5 s between CS and US. In effect, the CS replaces the US in eliciting the reflex response. The mechanism for classical conditioning is believed to be similar to that underlying sensitization. The depolarization of cell membranes in the synapse causes an inflow of Ca^{++}, which brings about transmitter release and postsynaptic response. During the conditioning process, the repeated temporal pairing of action potential (CS) with synaptic input (US) leads to a greater inflow of Ca^{++} with each action potential, leading to more transmitter release and greater synaptic response [286].

11.4   Central pattern generators

Much work on small neural systems has concentrated on rhythmic behaviours. The circuits which govern such behaviours in invertebrates are usually called *central pattern generators* (CPGs). A CPG is a neural circuit that can be activated to generate a motor pattern even when separated from its nervous system and deprived of any sensory feedback. The pattern produced in an intact organism will be somewhat different because of neuronal modulation from sensory input, but isolated CPGs still provide a starting point from which to understand rhythmic behaviours such as locomotion. Typical examples of these circuits are the network underlying escape swimming in the mollusc *Tritonia diomedea* and the CPG circuit found in the lobster stomatogastric ganglion.

When touched by a predator, the sea slug *Tritonia* escapes by making a series of between two and 20 alternating dorsal and ventral flexions. The burst pattern underlying this behaviour is produced by a CPG network made up of at

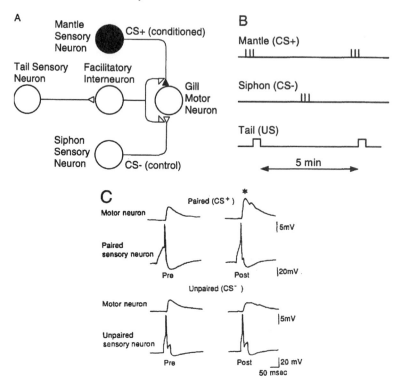

Figure 11.3. Classical conditioning of the gill-withdrawal reflex in *Aplysia*. (A) The neural system controlling the gill reflex: the US consists of strong electrical shocks to the sensory neurons in the tail; it always elicits a gill withdrawal. The CS+ consists of weak shocks to the sensory neurons of the mantle; by itself this does not elicit a response. Weak shocks to the sensory neurons of the siphon serve as a control (CS−). (B) The classical conditioning paradigm. CS+ is temporally paired with US; CS− is unpaired. (C) Intracellular recordings before conditioning (pre) and one hour after (post). The paired (CS+) trials produce a potentiation of the EPSPs in the motor neuron (marked ∗) whereas the unpaired (CS−) trials do not (adapted from [286]).

least 14 interneurons, each of which synapses onto motor neurons to produce the motor output [119].

The stomatogastric ganglion CPGs in the lobster controls the rhythmic movements of the teeth and the stomach muscles. One of the benefits of studying this network is that all of the neurons involved have been identified, and their connectivity is well understood [277]. Like CPGs in general, the network is dominated by mutually inhibiting neurons. This is a simple mechanism by which bursting neurons involved in rhythmic behaviours are kept synchronized. Electrically coupled neurons also play an important part in this coordination.

However, not all neurons that produce bursting behaviour are intrinsically bursting neurons. In the pyloric CPG, several motor neurons directly innervate the pyloric muscles; these are normally non-bursting, but can produce a single, long burst (or plateau potential) in response to a brief depolarization. The ability to burst in this way stems from the hormonal modulation which occurs via neurotransmitters from other ganglia. Thus, the neurons must be 'switched on' before they can burst. Bursting neurons can excite the more passive motor neurons into bursts of firing, with these bursts coordinated by mutual inhibition.

What is particularly interesting about the stomatogastric CPGs, however, is that bursting behaviour can be switched on selectively. In addition to supplying sensory feedback, input from other ganglia can create different patterns of activity in the CPG. Various transmitter substances can have different effects on each of the neurons in the circuit (switching them on or off or otherwise modulating their output), effectively 'rewiring' the network into different configurations [173].

Having outlined something about the links between neurophysiology and behaviour, and thereby illustrated implicitly the extreme simplifications made by the PDP style of modelling, we are now in a position to describe the 'computational neuroscience' paradigm which underpins our modelling work.

11.5   The computational neuroscience paradigm

During the period of the parallel distributed processing revival, there were similar radical upheavals happening in neurophysiology. To quote Thompson [313]:

> In the past generation, understanding of the biological basis of learning
> and memory has undergone a revolution. It now seems possible
> to identify the circuits and networks that participate in learning and
> memory, localise the sites of memory storage, and analyse the cellular
> and molecular mechanisms of memory.

As the 'explosion of discoveries over the last several decades concerning the structure of the brain at the cellular and molecular levels' [275] has progressed, there have been many models (both theoretical and computational) which attempt to explain the ramifications of the latest physiological discoveries.

Typically, such models aim for high physiological accuracy and deal with aspects like cable theory and compartmental models of neurons [318, 253, 273], ion-channels in single neurons [58, 183, 335], and synaptic transmission [222, 274] . However, interesting and potentially useful as these models are, they are concerned primarily with biochemistry and neurophysiology rather than intelligence and behaviour; they are tools for the physiologist rather than the cognitive scientist. Although it is apparently the case that brain biochemistry causes behaviour, if we are trying to understand the latter then we need to model at a level where we can keep both in sight, so we can understand how a particular biochemistry gives rise to a particular behaviour. As Sejnowski and his colleagues [275] have said on the subject of 'realistic' brain models:

> While this approach can be very useful, the realism of the model is both a weakness and a strength. As the model is made increasingly realistic by adding more variables and more parameters, the danger is that the simulation ends up as poorly understood as the nervous system itself.

Indeed, the concern of Sejnowski *et al* is to champion a new approach which they term 'computational neuroscience' and characterize as follows:

> The ultimate aim ... is to explain how electrical and chemical signals are used in the brain to represent and process information ... more is known now about the brain because of advances in neuroscience, more computing power is available for performing realistic simulations of neural systems, and new insights are available from the study of simplifying models of large networks of neurons.

Although neurophysiological understanding of neuronal networks is far from complete, we believe that enough is now known of their properties that we can make some headway with models built using the data we have. This sentiment is shared by many researchers. For instance, Koch and Segev [164] write:

> ... research in neuroscience has accumulated a wealth of experimental data about the brain ... our computers have given us the power to simulate and thereby understand some complex systems in great detail. Conjointly, these developments have led to the emergence of a new paradigm, a new approach toward understanding the nervous system: computational neuroscience.

Researchers in this new paradigm seem to be neurophysiologists who have discovered the power which computational modelling offers them, and PDP researchers who have despaired at the lack of realism in the connectionist paradigm. Selverston [276] describes PDP style neural networks as being typified by 'neurons with grossly over-simplified physiological properties, and synapses whose main function seems to be blind obedience to Hebbian learning rules'. The work of people such as Barish [31] on ion-channels as the basis for behavioural diversity and Llinas [178] on the variety of neurons in the mammalian central nervous system gives some idea as to the range and complexity of real neurons. (Llinas lists over 100 neuron types, along with the unique set of activators, inhibitors, thresholds, etc for each of them.) Crick [96] and Churchland and Churchland [68] have argued that neural net researchers would do well to take a closer look at real neurophysiology.

More and more models are being produced which fit the framework of computational neuroscience, linking some of the principles of neural nets with data from experimental neurophysiology. Examples of this kind include the work of: Gelperin *et al* [118] and Hopfield [145] on learning in the mollusc *Limax*; Selverston [276] and Kleinfield and Sompolinsky [162] on invertebrate central pattern generator circuits; Arbib [13] on visually guided behaviour in the frog; Getting [119] on the network underlying escape swimming of the mollusc *Tritonia*; Cliff [72] on visual processing in the hoverfly *Syritta*; Card and Moore [60] on the use of analogue CMOS circuits to simulate learning in *Aplysia* together with our own simulation [272] based on object-oriented programming (OOP); and Beer, Chiel and Stirling's model [38] of cockroach locomotion. In order to give a feel for the style of computational neuroscience, we briefly summarize the latter study.

The work of Beer *et al* shares with Cliff's the use of what are essentially PDP neuron models. However, the Beer model does include pacemaker neurons, and is based on architectures developed from neurophysiological studies. A circuit is developed for locomotion in the cockroach which features an identical six-neuron sub-circuit for each leg. The pattern generators on each leg exhibit phase-locking, through which particular gaits can be observed in the simulated animal. These gaits vary according to the animal's speed, and correspond exactly to those exhibited by the real creature. Even more intriguing, if the simulated animal is 'decapitated' and pushed along, the legs will still go through the same sequence of motions as when they were controlled by the animal, and will produce different gaits depending on the speed of pushing. This is not only an excellent example of independent subsystems being able to produce correct behaviour without any overall control (*à la Brooks*), but also again matches with the observed behaviour of the real cockroach in a similar headless condition. In our view, this is an exemplary piece of synthetic psychology. Although it is based on a real neural system, it is possible (although not attempted here) to see what each neuron is doing and what its function is.

There are aspects from all of these 'computational neuroscience' models which are very attractive, and which have been used in our own work.

11.6 *Hi-NOON* neural simulator

A program has been developed to assist the modelling of complex aspects of real neural systems (such as spiking, spontaneous firing, adaptation and synapse-on-synapse connections), while retaining much of the simplicity and elegance of PDP models. This section briefly describes the program, called *Hi-NOON* (hierarchical network of object oriented neurons). The program dates back several years [272] and, as Murre [201] points out: 'The field of neurosimulators is moving very fast', so that there are almost certainly other neural models now in existence which use similar techniques to greater effect.

In an attempt to exploit the best aspects of each, the program functions at a level between the two extremes of PDP and highly accurate neuronal modelling: individual neurons are considered at the level of membrane potential. This allows

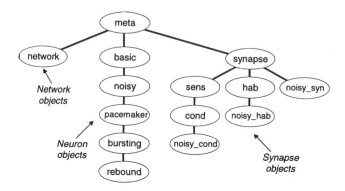

Figure 11.4. Hierarchy of objects in the network program. All objects are descended from the *meta* object.

outputs from the model to be compared directly with the wealth of physiological data obtained in intracellular recording so as to ensure physiological realism, but it also allows (within the limits of computer resources) reasonable numbers of neurons to be connected together and their network behaviour simulated. The methodology reflects our belief that the power of biological neural systems as information processors most likely arises as a product of complexity at both the neuron and the network level. We do not agree with the speculation of Rumelhart *et al* ([344], p 131) that 'individual neurons probably don't compute very complicated functions'. Learning is by biologically motivated mechanisms rather than biologically implausible supervised back-propagation.

An object-oriented programming (OOP) language is used, with networks, neurons, and synapses all represented as objects in a hierarchy (figure 11.4). The benefits of OOP are twofold. First, it allows polymorphism, so that a single network may contain many different types of neuron functioning at different levels of complexity. For instance, we can include PDP type units whose 'activation' values correspond roughly to the average spike rate. Second, the ability for objects to inherit properties from other objects makes it easy to define more physiologically exact neurons in terms of simpler neurons. Together, these allow us to simulate behaviour at the most appropriate level of abstraction in any given case. One disadvantage, however, is that OOP poses difficulties for creating the network itself, and setting parameters for the neurons and synapses, since at the top level of the hierarchy, the network itself is a single, complex object. To overcome this, a graphical 'click and drag' user interface (*NoonCAD*) has been added to facilitate the design of the neuronal circuit.

Hi-NOON is *phenomenological*, i.e. we model the effects of the processes going on in the neuron instead of the processes themselves. In other words, we code for function rather than mechanism. It differs from connectionist models in that it treats the neuron not as one big 'black box', but as several smaller ones.

From basic neurophysiology, it is fairly easy to see exactly which attributes a model neuron must have. The program uses a flexible parameter system to cater for differences between neurons, and to keep track of the changing state of a neuron over time. Most of the parameters have obvious correlates in the real neuron, the others are mainly 'book-keeping' parameters. Such a system allows circuits modelled on real data to be constructed with a minimum of effort.

11.6.1 *Network* objects

At present, the *network* is extremely simple. Its counter simply measures the time since the simulation began, and stops when this reaches the specified length of simulation. The method *update* in this case merely tells each of the network's neurons to update themselves. It does this via the method *read-level*, which returns the value of the object's *dynamic-level* after updating. Because the neurons themselves tell their synapses to update (as well as telling them to fire), a single call to the *network* object sets the whole network into action.

11.6.2 *Neuron* objects

Each of the *neuron* objects in the hierarchy adds complexity to its predecessor. Although we can consider neurons at many levels of complexity, typically we have used a state system to model the internal processes of the neuron. In this work, a neuron is treated as being in one of a number of (six) states depending on the present membrane potential, cell threshold and whether or not the cell has just fired. The use of a state system (described below) for controlling the membrane potential facilitates the addition of new features to the program; it is necessary only to identify which of the states may trigger this feature and to add a method call at that particular state. This, coupled with OOP's inheritance, allows models to be developed and altered relatively easily.

The neurons can be characterized as follows:

- This neuron introduces the state-system (overriding the *meta* object), and tells its synapses to fire when it is above threshold.
- This is similar to the basic neuron except that it also has a parameter to determine the amount of stochastic noise in the neuron. If this has a high value, the neuron may sporadically fire spontaneously. Noise can be turned off by setting this parameter to zero.
- In addition to the properties of the noisy neuron, the pacemaker fires at regular intervals.
- This uses the mechanisms of the pacemaker, but instead of firing a single spike, the neuron generates a short high-frequency burst of spikes.
- This uses some of the mechanisms of the bursting neuron in a slightly different way, to give a neuron which exhibits post-inhibitory rebound. This often means that if the membrane potential is forced above threshold, it will fire a train of spikes rather than a single spike.

The particular choice of neuron parameters is detailed in a previous paper [272]. Only the most complex of neurons (e.g. bursting) will have all of these parameters. To a certain extent, our choice aligns with that of Selverston [276].

11.6.3 *Synapse* objects

In *Hi-NOON*, synapses also have a set of parameters (also detailed in [272]). As with the neurons, only the most complex synapses will have all of the parameters set.

11.6.4 State-system algorithm

All of the *neuron* objects, from the basic object onwards, use a state system for simulating the changes in membrane potential. A neuron is treated as being in one (or, occasionally, more than one) of six states, depending on whether the cell has just fired, the present membrane potential, the threshold etc. For example, if the membrane potential (*dynamic-level*) is above the threshold, and the cell has not just fired, then the neuron will start to generate a spike and will initiate synaptic transmission. The use of a state system for controlling the membrane potential, rather than a series of differential equations, makes it much easier to add new features to the program.

11.6.5 Network learning

Habituation is programmed into the neuronal circuit simulation by the three parameters *base-level*, *dynamic-level*, and *recovery*. The parameter *base-level* is the 'normal' weight of the synapse and is a constant. The parameter *dynamic-level* holds the present synaptic strength and is variable during the program's run, and *recovery* is a constant (within each synapse) which determines how quickly *dynamic-level* returns to *base-level*. If a synapse is of a habituating type, then the *dynamic-level* of the synapse is lowered every time it is fired, but this effect diminishes with time (depending on the value of recovery).

The *sens* synapse uses the same parameters as are used in the habituation routines in order to simulate the Ca^{++} influx associated with sensitization (also known as presynaptic facilitation). The major difference is one of addressing. In the case of habituation, if synapse i of neuron A fires (synapse A_i), it transmits a signal to neuron B and the weight of A_i is changed. With sensitization, however, if A is a facilitatory interneuron, then the synapse A_i will not connect onto a neuron, but onto another synapse, B_j. When synapse A_i fires, it increases the *dynamic-level* of B_j. The program works out the pointer for this 'indirect' addressing when the network file is loaded and the objects constructed.

Conditioning uses a similar routine to sensitization, but the strength of the facilitation depends on the length of time since the target synapse last fired. The facilitatory effect is strongest when this period is approximately half a second.

This depends on the value of the target synapse's counter. As with habituation, the value of recovery (of the target synapse, in this case) determines how quickly the effects of sensitization or conditioning wear off.

11.7 Simulation of behaviour with *Hi-NOON*

Our thesis in this chapter is that neurophysiology provides a starting point for synthetic psychology, with particular reference to the design of adaptive mobile robots. In the previous section, we described the *Hi-NOON* neural simulator, whose purpose is to facilitate the design of a suitable 'nervous system' for the robot. Hence, it is necessary to confirm that *Hi-NOON* is indeed capable of simulating neurobiological behaviour appropriately.

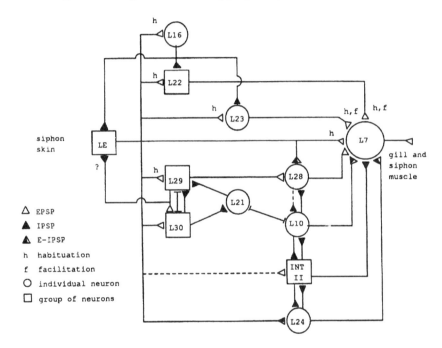

Figure 11.5. Summary diagram of the neural circuitry involved in the gill-withdrawal reflex of *Aplysia* (modified from [131]). Dotted lines indicate connections that may be indirect. The question mark by the L29–LE synapse indicates that the sign of the synaptic action is not known.

To this end, the simulator has been tested against a number of extant neural models of behaviour. In this section, we describe just one of these simulations: namely, the well studied gill-withdrawal conditioned reflex in *Aplysia*, whose neural circuitry is shown in figure 11.5. First, however, we demonstrate *Hi-NOON*'s ability to reproduce habituation and sensitization behaviour.

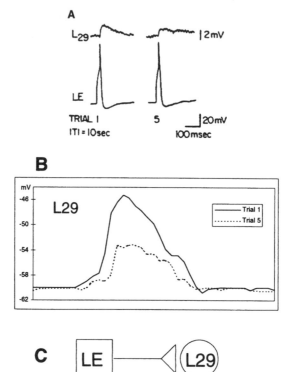

Figure 11.6. Habituation of the LE–L29 synapse. (A) Decrement of EPSP from an LE sensory neuron to L29. Action potentials in an LE neuron produce an EPSP in L29. The first and fifth PSPs are shown from a series of trials in which stimulation was repeated at 10 second intervals. (B) The model shows a similar habituation, although the inter-trial interval was only five seconds. (C) Diagram of the LE–L29 connection.

Figure 11.6 shows how the strength of the LE-L29 synapse decreases with repeated firing, illustrating the phenomenon of habituation. Output from the model shows a similar habituation in L29's responses. The presynaptic spike generated by the model LE neuron was the same for both trials.

Real neurophysiological data for actual neural circuits displaying sensitization seem hard to come by. Hence, the example given here is not based on any particular neural system. Instead, a prototypical three-neuron circuit has been used to show how a facilitatory interneuron making a synapse-on-synapse connection can increase the effective weight between two other neurons (figure 11.7).

In this example, neuron 1 is a noisy pacemaker, neuron 2 is a low-frequency pacemaker with a sensitizing synapse, and neuron 3 is a normally quiescent cell which receives excitatory postsynaptic potential (EPSPs) from neuron 1. Every

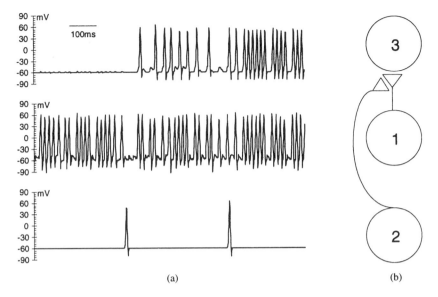

(a)   (b)

Figure 11.7. Computational model of sensitization. (a) Neuron 1 is a pacemaker cell with a weak synaptic connection to neuron 3. Neuron 2 is a facilitatory interneuron with a synapse-on-synapse connection onto the synapse between neurons 1 and 3. Every time neuron 2 fires, it enhances this synaptic connection. Eventually the pacemaker firing of neuron 1 produces one-for-one action potentials in neuron 3. (b) Diagram showing hypothetical interconnections of these neurons.

time neuron 2 fires, it facilitates the synapse from neuron 1 to neuron 3. It can be seen that the original sub-threshold responses of neuron 3 to neuron 1 become one-for-one action potentials.

Finally, figure 11.8 shows the model demonstrating classical conditioning as exhibited in *Aplysia*'s gill-withdrawal reflex. It shows both the neuron responses in the real circuit, and the output produced by *Hi-NOON* for the motor neurons under the same training conditions. Notice the increased EPSP in the paired condition.

Although the model exhibits the various simple forms of learning found in invertebrate systems, it also has the ability to display pacemaker and bursting behaviour. Such neuronal mechanisms are vital for the function of the rhythmic behaviours associated with invertebrate motor activity and central pattern generators. Space precludes a treatment of these matters here.

11.8 Adaptive robot behaviour

As an example of how we can go forward from neurophysiology to synthesize systems with interesting behaviour, we have used the *Hi-NOON* network

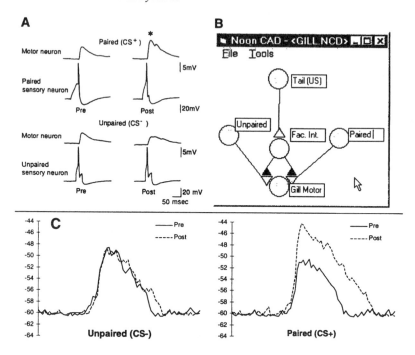

Figure 11.8. Computational model of classic conditioning. (A) Reaction of motor neuron to unpaired (CS−) and paired (CS+) stimulation before (pre) and after (post) a series of three trials in which the CS+ was paired with an unconditioned stimulus (electrical shock). (B) *NoonCAD* model of the gill-withdrawal network. Compare to the diagram of the real circuit given in figure 11.5. (C) *Hi-Noon* displays a similar conditioning. For clarity, only the postsynaptic (motor) neuron is shown.

simulator described in section 11.6 to consider Braitenberg vehicles in terms of real neurons. Ideally, we would like to build these vehicles and study their behaviour in the real world, and we have in fact done precisely this for vehicles with a fixed, non-adaptive nervous system. When it comes to adaptive behaviour, however, it is difficult to implement the desired learning behaviour in real time, either entirely on-board the vehicle or via an umbilical. Hence, the work described in this section simulates the vehicle's 'world'.

11.8.1   The five-neuron 'trick'

The simple Braitenberg vehicles described in the introduction contain (and therefore can be modelled using) just four neurons (two sensory and two motor). But our experience with modelling learning in *Aplysia* using *Hi-NOON* led us to ask the question of how the *hard-wired* architecture of such a vehicle could be altered so as to produce a *soft-wired*, adaptive version, which learned how to

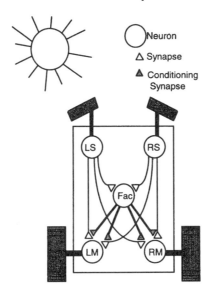

Figure 11.9. The five-neuron 'trick'—the two sensory neurons (LS & RS) have connections to both of the motor neurons (LM & RM), and also connect to the facilitatory interneuron (Fac). This in turn makes (conditioning) synapse-on-synapse connections to the four (habituating) synapses between sensory and motor neurons. This vehicle *adapts* in a way that makes it light seeking.

use its wheels so as to steer towards light. The answer is that a single additional, facilitatory interneuron, making synapse-on-synapse connections to the synapses from sensors to motors, allows the vehicle to learn (via classical conditioning) this behaviour (figure 11.9).

To explain: the vehicle is initially connected up completely neutrally, with connections of equal strength going from each sensory neuron to both of the motor neurons. These are 'noisy habituating' synapses which have a probability of firing proportional to the membrane potential of their parent neuron, and this weight decreases very slightly each time they fire. The facilitatory interneuron makes conditioning synapse-on-synapse connections with all four of the sensory to motor synapses. Now, in the figure the light is on the vehicle's left. Thus, the two synapses made by the LS neuron are more likely to fire than the ones made by the RS neuron. The two synapses made by the left sensor are both equally likely to fire. However, because of their 'noisy' nature they will not necessarily fire at the same time. If the left-hand synapse fires (the straight connection), then the vehicle will turn away from the light. This means that the facilitatory neuron (which receives input from both light sensors) is less likely to fire, and the strength of the left-hand synapse will not be increased. In fact, it will decrease slightly because of the effects of habituation. If, on the other hand, the right-hand synapse fires (the crossed connection), then the vehicle will turn towards the light, the facilitatory neuron (now receiving more input from the two sensory neurons) is more likely to fire, and the strength of the right-hand synapse will increase as a result of classical conditioning (which greatly outweighs the slight habituation).

The system as a whole is riddled with noise, and there is a delicate balance between the antagonistic effects of habituation and conditioning. However, a vehicle originally wired up neutrally (it will move faster in lighter areas, but the direction of movement is essentially random) will gradually adapt so that its crossed connections become stronger (as a result of conditioning), while the straight connections become weaker (because of habituation). The five-neuron vehicle therefore effectively builds itself the required circuit (figure 11.10(a)).

This system has been tested in a simple simulated environment of a single light source and two obstacles. Figure 11.10(b) shows the path of a vehicle which has only been 'alive' for a few minutes. It is clear that already some adaptation has taken place. Figure 11.10(c) depicts the behaviour of a vehicle which has already been 'in the world' for around 20 minutes and has thus better developed its light-seeking behaviour.

During the running of these simulations, it became clear that nearly all of the adaptation was taking place at the borders between light and shadow. This is because at these boundaries turning one way (towards the light) will result in reinforcement while turning the other (into the shadow) results in none. In fact, the vehicle never learns in an environment where there are no objects, indicating that a 'complex' environment is a help rather than a hindrance. The five-neuron vehicle is a soft-wired version of Braitenberg's original light-seeking (called 2b by him) vehicle. Its 'goal' is the same, but its nervous system adapts to achieve this goal. Even if we were to take the two light sensors and cross them over, the vehicle would still (eventually) orientate towards the light.

This suggests that the benefits brought by synaptic plasticity may come fairly cheaply (in this case, by the addition of a single neuron). Indeed, recent research on insect vision has shown that such plasticity is far more common in invertebrates than had been previously thought (e.g. [334]). The vehicle's nervous system has no explicit knowledge built in as to how to move towards light, yet by its very nature the circuit is light seeking. This not only shows the potential power of a five-neuron circuit, but also illustrates how just a few cellular 'building blocks' can be combined to produce a complex behaviour.

We emphasize that the conditioning taking place here results from the synapse-on-synapse connections. This places the model at a much lower level than that normally employed in autonomous vehicles that learn via conditioning. Good examples of higher-level conditioning architectures are the use of Klopf *et al* of a hierarchical network of control systems [163] (which displays a level of associative chaining), and Sutton's system [302] which uses reinforcement learning based on a comparison of real world and internal model. It would be an interesting task to try and combine the above-described neural model with

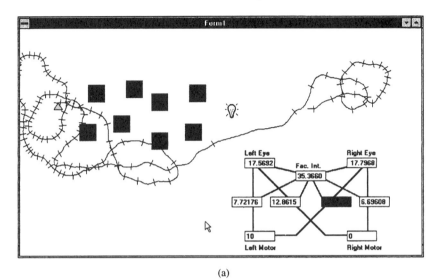

(a)

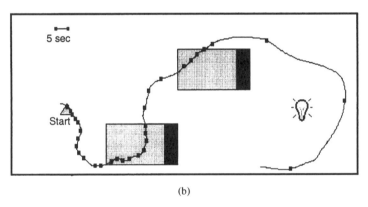

(b)

Figure 11.10. (a) A five-neuron vehicle learns to adapt to an environment of light and shadow (grey squares). The network in the lower right corner shows neuron activations and connection weights. The connection between the right eye and the left motor has just fired. (b) After training for a couple of minutes, a five-neuron vehicle exhibits signs of light-seeking. Shaded areas show shadows behind objects. (c) A vehicle which has already been adapting for some time shows light-seeking and dark-avoidance behaviour.

these higher-level, more symbolic architectures. Indeed, it is worth noting Sutton and Barto's [303] observation that the delta rule (of which back-propagation is a generalization) is nearly identical in form to Rescorla and Wagner's [256] model of classical conditioning. (We have presented back-propagation and biological forms of learning, such as conditioning, as essentially orthogonal and conflicting—as indeed they are at a low level of description. Thus, the

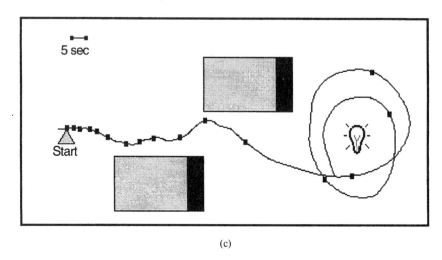

(c)

Figure 11.10. Continued.

convergence of high-level behaviours is interesting.) Schmajuk [270] provides a useful overview of the different types of computational model that have been used to describe classical and operant conditioning, while Sharkey (this volume) has more to say about robot training methods in general.

11.8.2   A more complex adaptive vehicle

While the five-neuron vehicle has a certain minimalist appeal, we wanted to show how we could combine several mechanisms from invertebrate neurophysiology in a single adaptive vehicle. The neural circuit involved is too large to be immediately understandable as a whole, so it is explained here in the same way as it was originally constructed; by adding adaptive mechanisms onto a hard-wired base (developed jointly with Adam Maddison, Department of Psychology, University of Nottingham).

11.8.2.1   The central pattern generator

At the heart of the circuit is a simple hard-wired central pattern generator (CPG). This provides the basic driving force for the vehicle, and produces a simple wandering behaviour. The CPG itself is a four-neuron loop with mutual inhibition between two of the neurons (figure 11.11(A)). Although not based on any specific invertebrate, this CPG uses principles from identified neural circuitry (e.g. the *Clione* swim network [268]).

There is an excitatory connection from neuron C1 to the forward motor, and from C2 and C4 to the right and left steer motor neurons respectively. The synapses to the steer neurons are of type 'noisy', which means that the

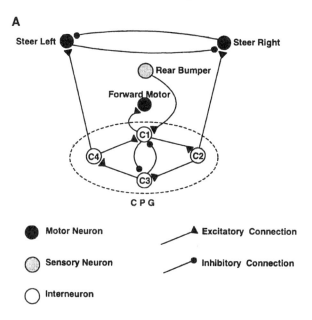

Figure 11.11. The hard-wired basic network. (A) The central pattern generator CPG consists of four neurons (C1–C4). The CPG is 'kickstarted' by the Rear Bumper sensory neuron, and gives rise to a basic wandering behaviour (described in detail in the text). The connections between C2 and Steer Right and between C4 and Steer Left are noisy synapses. (B) The basic reflex behaviour. Left and Right Bumper sensory neurons are connected to the Left and Right Reflex interneurons respectively. Each Reflex interneuron is connected to the Reverse Motor neuron and the ipsilateral Steer neuron. This causes the robot to reverse while steering toward the object that it hit before resuming its wandering behaviour.

probability of the synapse firing depends on the activity of the parent neuron, thus allowing for deviation from the robot's generally straight-line path when wandering over long distances.

The CPG does not start firing spontaneously; an excitatory connection from the rear bumper sensory neuron activates C1, thus starting the reverberatory pattern of firing of the CPG. (This is a practical solution that allows the robot to be 'started' by pressing the back bumper.) As the C2 and C4 neurons make connections to the right and left steer motors, a 'swimming' side-to-side motion results when the robot is moving forward as a result of activity of the CPG.

11.8.2.2 *Reflex responses*

The design of the reflex responses built into the neural circuit is based upon the simple monosynaptic reflex responses found in all animals. It is these

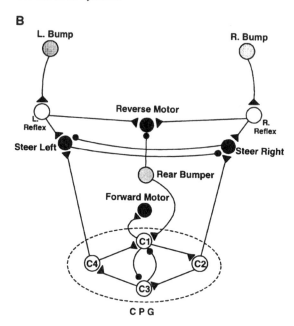

Figure 11.11. Continued.

reflex responses that form the starting point for learning in the robot. The network has two hard-wired reflex responses that are used as the unconditioned stimuli for classical conditioning. Each of the front bump sensors makes an excitatory connection to the ipsilateral reflex interneuron. These reflex neurons in turn connect to the reverse motor neuron and the ipsilateral steer motor (figure 11.11(B)).

The design is such that if, for example, the left bump sensor is activated then the robot will reverse away from the object that it collided with while steering towards it. This has the effect that once the reflex has been completed the robot will no longer be facing toward the object with which it collided. This has an obvious analogy with the 'withdraw-from-pain' reflex that all animals possess. Although the effect is hard-wired (and there are good evolutionary reasons why it should be), it can act as the basis for adaptive behaviour.

There is one final reflex that plays no part in the adaptation of the robot. An inhibitory connection from the rear bumper sensory neuron to the reverse motor neuron means that the vehicle will stop reversing if it backs into something, allowing the robot to wander forward again under the control of the CPG.

11.8.2.3 Classical conditioning

The neural mechanisms of classical conditioning used in this circuit are exactly the same as those described earlier in this chapter. Each bump sensory neuron

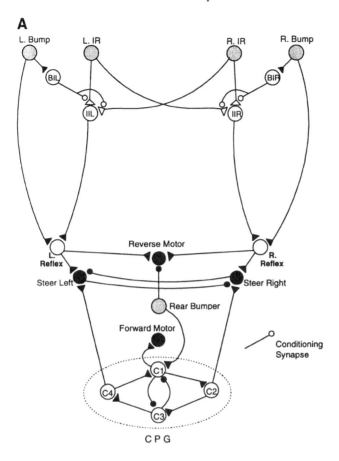

Figure 11.12. Adding adaptation to the network. (A) Left and Right Bump Interneurons (BIL and BIR) make conditioning synapse-on-synapse connections with the synapses from the Infrared sensory neurons (L. IR and R. IR) to the Left and Right Infrared Interneurons (IIL and IIR) respectively. These interneurons make connections to the Reflex neurons. (B) Mechanisms for detecting changes in light. Each of the light sensory neurons (L. LDR and R. LDR) makes excitatory connections to a delay interneuron and the $\Delta+$ neuron, and an inhibitory connection to the $\Delta-$ neuron. The delay neuron in turn makes an excitatory connection to $\Delta-$ and an inhibitory one to $\Delta+$. The effect is that the $\Delta+$ neuron responds to an increase in light level while the $\Delta-$ neuron becomes active when the light decreases.

makes an excitatory connection to its facilitatory interneuron (BIL and BIR) which, in turn, makes conditioning synapse-on-synapse connections to the synapses connecting (from the infrared sensory neurons) to the ipsilateral infrared interneuron (IIL and IIR, respectively). Each infrared sensory neuron (IR) makes connections to both of the infrared interneurons (see figure 11.12(A)): thus,

B

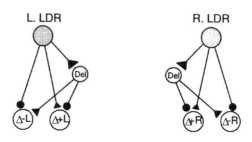

Figure 11.12. Continued.

the network is initially wired up neutrally. The connections between the IR sensory and interneurons are very weak initially; even if both sensory neurons fire, the IR interneurons will not. This means that, to begin with, the robot effectively ignores the signals from the IR sensory neurons—they have no effect on its behaviour. However, as the robot explores its world, the connections from the IR sensory neurons become reinforced in such a way that the robot uses the information from its IR neurons to avoid hitting obstacles. (In fact, IR signals lead to reflex behaviour as if the robot had struck an obstacle.) In order to see how this is achieved, we need to engage in a small thought experiment.

Imagine that the robot is wandering around when it approaches an obstacle on its left-hand side. As it gets to within a few inches of the obstacle, the left IR sensory neuron begins to fire (because an IR sensor is active when it picks up IR signals generated by the robot being reflected back off nearby objects). However, although the IR sensory neuron sends signals to both IR interneurons (IIL and IIR), the connection is too weak for the interneurons to fire, so the robot continues forward. A second or two later, the left bump sensor hits the object. The immediate effect is that the left reflex interneuron is triggered and the robot begins its reflexive withdrawal behaviour. The other effect is that the bump interneuron (BIL) fires, leading to possible conditioning of the connections from the two IR sensory neurons to IIL. Only the left IR sensory neuron has fired recently, so the connection from that neuron to IIL will be facilitated—its weight will be increased. In future (or perhaps after one or two more 'trials' depending on the exact parameters used), the connection from the left IR sensory neuron and the IIL is strong enough that the IR signal on its own is enough to trigger the left reflex interneuron and cause the backing off behaviour. The robot has learned to use a different sensory modality to avoid the 'pain' of colliding with obstacles—as the robot gets to within a few centimetres of an object, its IR fires and it reverses away.

Once these conditioned connections have been formed, the next stage of adaptation takes place.

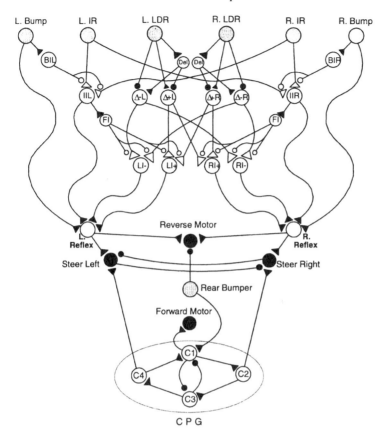

Figure 11.13. The complete network. The Infrared interneurons (IIL and IIR) each excite a Facilitatory Interneuron (FI). Each FI makes conditioning synapse-on-synapse connections with the synapses from the Light Δ+ and Δ− neurons to the ipsilateral Light Interneurons (LI+, LI−, RI+, and RI−). These, in turn, make connections to the Reflex interneurons. The network as a whole displays wandering behaviour, and the vehicle backs away when it hits an object. The network adapts at first to back away from infrared signals as it learns that they tend to precede contact with an obstacle. Once this learned behaviour is established, the system adapts so that it backs away from decreases in light because shadows tend to precede infrared signals (and therefore obstacles).

11.8.2.4 Chained conditioning

The task of learning to avoid obstacles through the association of objects with shadows occurs in a similar manner to the association of bump and IR detection previously described. However, the light sensors (LDRs) generate a signal between 0 and 255 rather than the 0 or 1 output of the bump sensors and IRs and we were also eager not to pre-judge the environment for the robot by simply having 'shadow detectors'. Because of this, we used some additional neural mechanisms (similar to the Reichardt [254] motion detector) as shown in figure 11.12(B). The LDR sensory neuron makes an inhibitory connection to the $\Delta-$ interneuron and excitatory connections to the $\Delta+$ interneuron and a 'delay' neuron. This delay neuron makes an excitatory connection to $\Delta-$ and an inhibitory one to $\Delta+$. This gives two neurons for each LDR—one $\Delta+$ reports an increase in light at that LDR while the other $\Delta-$ reports a decrease. In the complete circuit (figure 11.13), the four Δ neurons make weak connections to the set of light interneurons (LI$-$, LI$+$, RI$+$ and RI$-$) so that each Δ neuron connects to the ipsi- and contralateral neuron with the same sign. These connections can be conditioned by the synapse-on-synapse connections from the facilitatory interneuron (FI) attached to the infrared interneuron, but initially are not strong enough to cause the light interneurons (and therefore the reflex) to fire.

As for the learning, we can use a similar thought experiment to the one used previously. The robot enters a patch of shadow behind an obstacle to its left. This has no immediate effect on the behaviour, but does cause the $\Delta-$L neuron to fire. A moment later, the left IR picks up the obstacle and fires, causing IIL to fire. This not only triggers the learned reflex behaviour, but also causes the left FI to fire, which strengthens the connection from a Δ neuron to the left light interneurons if that Δ neuron has recently fired. In this case, $\Delta-$L has recently fired so that connection is strengthened. It typically takes two to three incidents such as this before the robot starts to back away from shadows even without the IR firing.

Notice that the initial wiring of the vehicle does not make assumptions about the world (except for the fact that touch is a shorter-range modality than IR which in turn is shorter range than light). For example, if we were to put reversing glasses on our robot, or if obstacles were bright rather than dark, the neural circuit would still adapt correctly to its environment.

11.9   Discussion and conclusions

In this chapter, we set out to show that the simple forms of learning present in invertebrate systems can be of great use in designing 'nervous systems' for situated autonomous robots. Any observer trying to analyse the behaviour of even the simplest 'creatures' (animal or robot) can be seriously misled as to the architecture of that creature. By examining structure rather than guessing at function, we can start to build up to a model of intelligent behaviour rather than trying to deconstruct patterns of behaviour into their constituent 'representations'. In other words, modelling neurophysiology is a bottom-up synthetic process rather than a top-down analytic process.

Having shown the sorts of problems synthetic vehicles create for the analytic observer, we went on to consider the best approach to modelling a synthetic nervous system, comparing the well established PDP paradigm with more biologically faithful models. It was argued that the forms of learning displayed by even a creature as lowly as *Aplysia* were more suited to the real world than the complex learning algorithms employed within the PDP paradigm. Indeed there is something faintly ridiculous about the image of a researcher following a mobile robot around the laboratory, pressing a button on its head to provide error feedback every time it makes the wrong move [207]!

We next reviewed the neurophysiology of learning before describing central pattern generators, mostly because this kind of neuron turns out to be useful in our modelling but also to illustrate the wide gulf between the simple, abstract, homogeneous processing nodes found in PDP networks and the complexity and differentiation of real neurons. A brief overview of the field of computational neuroscience was then presented so as to motivate our description of the *Hi-NOON* simulator. *Hi-NOON* is a program which models neurons at a level of abstraction suited to the sorts of behaviour we would like to see in a situated system, while the subsidiary program *NoonCAD* facilitates construction and 'design' of neural systems. The simulator uses a level of abstraction between that of connectionist models and that of the more accurate neurophysiological models typically used to model single cells. This allows us to understand behaviour at both a neuronal and a network level and clarifies the relationship between the two. It is achieved by modelling the effects of the processes occurring in the neuron instead of modelling the processes themselves—in other words, coding for function rather than mechanism. To illustrate the virtues of the approach, we briefly described a computational model of the gill-withdrawal reflex of *Aplysia* based on *Hi-NOON*. This model was shown to exhibit the important forms of non-associative and associative biological learning.

We then presented two specific examples of the application of these concepts to adaptive robot behaviour. The first—which we call the five-neuron trick—shows how the light-seeking behaviour of a type 2b Braitenberg vehicle can emerge as an *a posteriori* consequence of experience of its world, rather than as a direct, *a priori* consequence of its rigid structure. In this way, the problem of how to give a system goals without resorting to external intervention has been solved. The 'goal' of the system is still determined by its initial wiring, but the way that this goal is achieved (and the effective wiring of the network) is determined by the manner in which the system adapts to its environment. The second, larger circuit shows how we can combine the driving force of a central pattern generator and hard-wired reflexes with classical conditioning so that our robot learns to use its various sense modalities to good effect. Virtually all animals have simple, hard-wired reflex behaviours such as withdrawing or escaping from pain and there are obviously good evolutionary reasons why this is so. Real animals can be conditioned so that they will display withdrawal behaviours to a generally harmless stimulus if that stimulus is consistently paired with an unpleasant sensation. Again, there are good reasons for this—if the black-and-yellow thing stings me, I would do well to avoid black-and-yellow things in future. Our robot is using the same neural mechanisms to achieve this result. By noticing that IR always precedes a bump ('pain') the robot can respond to the IR signal rather than the bump itself, which obviously has evolutionary value.

By taking principles from invertebrate neurophysiology and synthesizing simple artificial networks which use these principles, we gain valuable insight into how to solve the 'bootstrapping' problem of intelligence in autonomous robots. In short, neurophysiology can help with building better mobile robots, and building mobile robots can help us to understand neurophysiology.

Acknowledgments

Some of this chapter was prepared whilst Robert Damper was with the Department of Computer Science and Engineering, Oregon Graduate Institute of Science and Technology, USA.

Bibliography

[1] A. Browne *Neural Network Analysis, Architectures and Algorithms* Institute of Physics Publishing, Bristol, UK, 1997.

[2] A. Caramazza and E. B. Zurif. Dissociation of algorithmic and heuristic processes in language comprehension: Evidence from aphasia. *Brain and Language*, 3:572–582, 1976.

[3] W. A. Cook *Case Grammar Theory*. Georgetown University Press, Washington, DC, 1989.

[4] M. Abeles, H. Bergman, E. Margalit and E. Vaadia. Spatiotemporal firing patterns in the frontal cortex of behaving monkeys. *Journal of Neurophysiology*, 70:1629–1638, 1993.

[5] D. H. Ackley, G. E. Hinton and T. J. Sejnowski. A learning algorithm for Boltzmann machines. *Cognitive Science*, 9:147–169, 1985.

[6] V. Ajjanagadde and L.Shastri. Efficient inference with multi-place predicates and variables in a connectionist system. In *Proceedings of the 11th Annual Cognitive Science Society Conference*, pages 396–403, Erlbaum, Hillsdale, NJ, 1989.

[7] F. Alavi and J. G. Taylor. A global competitive neural network. *Biological Cybernetics*, 72:233–248, 1995.

[8] I. Aleksander and P. Burnett. *Reinventing Man*. Penguin, Middlesex, 1984.

[9] D. Alkon. *Memory Traces in the Brain*. Cambridge University Press, Cambridge, UK, 1987.

[10] P. Anandan, S. Letovsky and E. Mjolness. Connectionist variable binding by optimization. In *Proceedings of the 11th Annual Conference of the Cognitive Science Society*, pages 388–395, 1989.

[11] J. Anderson and G. Bower. *Human Associative Memory*. Winston, Washington, DC, 1973.

[12] J. R. Anderson. *The Architecture of Cognition*. Harvard University Press, Cambridge, MA, 1983.

[13] M. A. Arbib. Levels of modeling of mechanisms of visually guided behavior. *Behavioral and Brain Sciences*, 10:407–465, 1987.

[14] R. Arkin. Reactive robotic systems. In M. Arbib, editor, *The Handbook of Brain Theory and Neural Networks*, pages 793–796, MIT Press, Cambridge, MA, 1995.

[15] J. J. Atick and A. N. Redlich. Towards a theory of early visual processing. *Neural Computation*, 2:308–320, 1990.

[16] J. J. Atick and A. N. Redlich. What does the retina know about natural scenes? *Neural Computation*, 4:196–210, 1992.

[17] J. J. Atick and A. N. Redlich. Convergent algorithm for sensory receptive field development. *Neural Computation*, 5:45–60, 1993.

[18] F. Attneave. Some informational aspects of visual perception. *Psychological Review*, 61:183–193, 1954.

[19] J. B. Pollack Recursive distributed representations. *Artificial Intelligence*, 46:77–105, 1990.

[20] A. Baddeley. Is working memory working? *Quarterly Journal of Experimental Psychology*, 44:1–31, 1992.

[21] P. Bak and K. Chen. Self-organized criticality. *Scientific American*, 264(1):46–53, 1991.

[22] P. Bak, K. Chen and C. Tang. A forest-fire model and some thoughts on turbulence. *Physics Letters*, 147A:297–300, 1990.

[23] P. Bak, C. Tang and K. Wiesenfeld. Self-organized criticality: An explanation of $1/f$ noise. *Physical Review Letters*, 59:381–384, 1987.

[24] T. E. Baker. *Implementation Limits for Artificial Neural Networks*. Master's thesis, Oregon Graduate Institute of Science and Technology, Oregon, 1990.

[25] P. Bakker and Y. Kuniyoshi. Robot see, robot do: An overview of robot imitation. In *Proceedings of the Workshop on Learning in Robots and Animals, AISB-96*, 1996.

[26] A. Baldwin. Subsymbolic inference: Inferring verb meaning. In *Proceedings of the 11th European Cybernetics and Systems Conference*, pages 105–109, 1992.

[27] D. H. Ballard. Parallel logical inference and energy minimisation. In *Proceedings of the AAAI National Conference on Artificial Intelligence*, pages 203–208, 1986.

[28] D. H. Ballard and C. M. Brown. *Computer Vision*. Prentice-Hall, Englewood Cliffs, NJ, 1982.

[29] A. Baloch and A. Waxman. Visual learning, adaptive expectation and behavioural conditioning of the mobile robot MAVIN. *Neural Networks*, 4(3):271–302, 1991.

[30] I. L. Balogh. *An analysis of Connectionist Internal Representation*. PhD thesis, New Mexico State University, Las Cruces, NM, 1994.

[31] M. E. Barish. Ion channels as a source of behavior: doing more with less in 'simpler' organisms. *Trends in Neuroscience*, 11:558–561, 1988.

[32] H. B. Barlow. Three points about lateral inhibition. In W. Rosenblith, editor, *Sensory Communication*, pages 782–786, MIT Press, Cambridge, MA, 1961.

[33] H. B. Barlow and P. Földiák. Adaptation and decorrelation in the cortex. In H. B. Barlow and P. Földiák, editors, *The Computing Neuron*, pages 54–72, Addison-Wesley, Wokingham, 1989.

[34] H. B. Barlow and J. D. Mollon. *The Senses*. Cambridge University Press, Cambridge, 1982.

[35] H. B. Barlow. Single units and sensation: A neuron doctrine for perceptual psychology? *Perception*, 1:377–394, 1972.

[36] J. Barnden and K. Srinivas. Overcoming rule-based rigidity and connectionist limitations through massively parallel case-based reasoning. *International Journal of Man–Machine Studies*, 36:221–246, 1992.

[37] S. Becker and G. E. Hinton. *Spatial Coherence as an Internal Teacher for a Neural Network. Technical Report CRG-TR-89-7*, Department of Computer Science, University of Toronto, December 1989.

[38] R. D. Beer, J. Chiel and L. S. Sterling. An artificial insect. *American Scientist*, 79:444–452, 1989.

[39] I. S. N. Berkeley, M. R. W. Dawson, D. A. Medler, D. P. Schopflocher and L. Hornsby. Density plots of hidden value unit activations reveal interpretable bands. *Connection Science*, 7(2):167–186, 1995.

[40] N. O. Bernsen and I. Ulboek. Two games in town: Systematicity in distributed connectionist systems. *AISB Quarterly*, 79:25–30, 1992.

[41] D. S. Blank, L. A. Meeden and J. B. Marshall. Exploring the symbolic/subsymbolic continuum: a case study of RAAM. In J. Dinsmore, editor, *The Symbolic and Connectionist Paradigms: Closing the Gap*, pages 113–148, Erlbaum, Hillsdale, NJ, 1992.

[42] M. Bodén. *Representing and Reasoning with Nonmonotonic Inheritance Structures using Context-sensitive Connectionist Representations*. PhD thesis, University of Exeter, 1996. In preparation.

[43] M. Bodén and L. Niklasson. Features of distributed representations for tree-structures: a study of RAAM. In L. Niklasson, editor, *Current Trends in Connectionism—Proceedings of the 1995 Swedish Conference on Connectionism*, pages 121–140, Erlbaum, Hillsdale, NJ, 1995.

[44] P. Bourgine and F. Varela. Towards a practice of autonomous systems. In F. Varela and P. Bourgine, editors, *Toward a Practice of Autonomous Systems*, pages xi–xvii, MIT Press, Cambridge, MA, 1992.

[45] V. Braitenberg. *Vehicles: Experiments in Synthetic Psychology*. MIT Press, Cambridge, MA, 1984.

[46] V. Braitenberg and A. Schüz. *Anatomy of the Cortex: Statistics and Geometry*. Springer, 1991.

[47] D. Broadbent. A question of levels—comments on McClelland and Rumelhart. *Journal of Experimental Psychology General*, 114(2):189–192, 1985.

[48] R. Brooks. A robust layered control system for a mobile robot. *IEEE Journal of Robotics and Automation*, RA-2:14–23, 1986.

[49] R. Brooks. Intelligence without reason. In *Proceedings of the International Joint Conference on Artificial Intelligence*, pages 569–595, 1991.

[50] R. A. Brooks. Intelligence without representation. *Artificial Intelligence*, 47:139–159, 1991.

[51] O. Brousse. *Generativity and Systematicity in Neural Network Combinatorial Learning. Technical Report CU-CS-676-93*, University of Boulder, 1993.

[52] G. Brown and C. Desforges. *Piaget's Theory: A Psychological Critique*. Routledge and Kegan Paul, 1979.

[53] A. Browne. Measuring distribution in distributed representations. *Neural Processing Letters*, 3:73–79, 1996.

[54] A. Browne and J. Pilkington. Performing structure-sensitive processes on sub-symbolic representations. In *Proceedings of the Irish Neural Networks Conference*, pages 19–24, 1994.

[55] A. Browne and J. Pilkington. Unification using a distributed representation. *Association for Computing Machinery Bulletin on Artificial Intelligence*, 5(2):5–7, 1994.

[56] A. Browne and J. Pilkington. Performing variable binding with a neural network. In J. G. Taylor, editor, *Neural Networks*, pages 71–84, Waller, Henley on Thames, 1995.

[57] B. Buchanan and E. Shortliffe. *Rule-Based Reasoning: the MYCIN Experiment.* Addison-Wesley, Reading, MA, 1984.

[58] J. H. Byrne. Comparative aspects of neural circuits for inking behavior and gill withdrawal in Aplysia californica. *Journal of Neurophysiology,* 45:98–106, 1980.

[59] S. C. Kwasny and B. L. Kalman. Tail-recursive distributed representations and simple recurrent networks. *Connection Science,* 7:61–80, 1995.

[60] H. C. Card and W. R. Moore. Silicon models of associative learning in Aplysia. *Neural Networks,* 3:333–346, 1990.

[61] N. R. Carlson. *Physiology of Behaviour.* Allyn and Bacon, Boston, MA, 1991.

[62] G. Chaitin. Randomness and mathematical proof. *Scientific American,* 233:47–52, 1975.

[63] D. J. Chalmers. Syntactic transformations on distributed representations. *Connection Science,* 2(1 and 2):53–62, 1990.

[64] P. Chen, K. Bak and M. H. Jensen. A deterministic critical forest fire model. *Physics Letters,* 149A:207–210, 1990.

[65] E. Cherniak. *Minimal Rationality.* Academic, San Diego, CA, 1986.

[66] N. Chomsky. Formal properties of grammars. In R. D. Luce, R. B. Bush and E. Galanter, editors, *Handbook of Mathematical Psychology,* volume 2, Wiley, New York, 1963.

[67] L. Chrisman. Learning recursive distributed representations for holistic computation. *Connection Science,* 3(4):345–366, 1991.

[68] P. M. Churchland and P. S. Churchland. Could a machine think? *Scientific American,* 262:26–31, 1990.

[69] A. Clark. Being there: Why implementation matters to cognitive science. *Artificial Intelligence Review,* 1:231–244, 1987.

[70] A. Clark. *Microcognition: Philosophy, Cognitive Science and Parallel Distributed Processing.* Bradford Books/MIT Press, Cambridge, MA, 1989.

[71] D. Cliff. Computational neuroethology: A provisional manifesto. In J. Meyer and S. Wilson, editors, *From Animals to Animats,* pages 29–39. MIT Press, Cambridge, MA, 1991.

[72] D. T. Cliff. The computational hoverfly: A study in computational neuroethology. In J-A. Meyer and S. W. Wilson, editors, *From Animals to Animats: Proceedings of the First International Conference on Simulation of Adaptive Behaviour,* pages 87–96, Bradford Books/MIT Press, Cambridge, MA, 1991.

[73] A. Collins and R. Michalski. The logic of plausible reasoning: A core theory. *Cognitive Science,* 13(1):1–49, 1989.

[74] G. W. Cottrell and J. Metcalfe. EMPATH: Face, emotion and gender recognition using holons. In *Neural Information Processing Systems 3.* Morgan Kaufmann, San Mateo, CA, 1990.

[75] J. J. Craig. *Introduction to Robotics.* Addison-Wesley, Reading, MA, 1986.

[76] F. Crick. The recent excitement about neural networks. *Nature,* 337:129–132, 1989.

[77] J. P. Crutchfield and K. Young. Inferring statistical complexity. *Physical Review Letters,* 63:105–108, 1989.

[78] J. P. Crutchfield and K. Young. Computation at the onset of chaos. In W. H. Zurek, editor, *Complexity, Entropy and the Physics of Information,* volume VIII. Addison-Wesley, Reading, MA, 1990.

[79] D. Servan-Schreiber, A. Cleeremans and J. L. McClelland. Graded state machines: The representation of temporal contingencies in simple recurrent networks. *Machine Learning*, 7:161–194, 1991.

[80] K. Dautenhahn. Getting to know each other—artificial social intelligence for autonomous robotics. *Robotics and Autonomous Systems*, 16:333–356, 1995.

[81] M. Davies and G. W. Humphreys. *Consciousness*. Blackwells, Oxford, UK, 1993.

[82] E. Davis. *Representations of Commonsense Knowledge*. Morgan Kaufmann, San Mateo, CA, 1990.

[83] D. C. Dennett. *Consciousness Explained*. Allan Lane, London, 1991.

[84] de Gennes. Phase transitions and turbulence: An introduction. In T. Riste, editor, *Fluctuations, Instabilities and Phase Transitions*, volume B11, NATO ASI Series. Plenum, New York, 1975.

[85] G. E. Hinton, D. E. Rumelhart and R. J. Williams. Learning internal representations by error propagation. In *Parallel Distributed Processing*, pages 533–536. Morgan Kauffman, San Mateo, CA, 1986.

[86] D. Dennett and M. Kinsbourne. Time and the observer. *Behavioural and Brain Sciences*, 15:183–247, 1992.

[87] M. Derthick. Mundane reasoning by parallel constraint satisfaction. *Technical Report TR CMU-CS-88-182*, Carnegie-Mellon University, 1988.

[88] A. K. Dewdney. Computer recreations—Braitenberg memoirs: vehicles for probing behavior. *Scientific American*, 256:8–12, 1987.

[89] T. J. Dietterich, H. Hild and G. Bakiri. A comparison of ID3 and backpropagation for English text-to-speech mapping. *Machine Learning*, 18, 1995.

[90] C. P. Dolan and P. Smolensky. Tensor product production system—a modular architecture and representation. *Connection Science*, 1(1):53–68, 1989.

[91] D. W. Dong and J. J. Atick. Temporal decorrelation: a theory of lagged and nonlagged responses in the lateral geniculate nucleus. *Network: Computation in Neural Systems*, 6:159–179, 1995.

[92] H. Dreyfus. *What Computers Can't Do*. Harper and Row, New York, 1972.

[93] H. Dreyfus and S. Dreyfus. *Mind Over Machine*, Free Press, New York, 1987.

[94] M. G. Dyer. Connectionist natural language processing: A status report. In R. Sun and L. A. Bookman, editors, *Computational Architectures Integrating Neural and Symbolic Processes*, pages 389–429, Kluwer, Boston, MA, 1995.

[95] N. E. Sharkey and A. J. C. Sharkey. A modular design for connectionist parsing. In M. F. J. Drossaers and A. Nijholt, editors, *Twente Workshop on Language Technology 3: Connectionism and Natural Language Processing*, pages 87–96, Enschede, 1992. Department of Computer Science, University of Twente.

[96] F. Crick. *The Astonishing Hypothesis*. Simon and Schuster, London, 1994.

[97] M. F. St. John. The story gestalt: A model of knowledge-intensive processes in text comprehension. *Cognitive Science*, 16:271–306, 1992.

[98] M. F. St. John and J. L. McClelland. Learning and applying contextual constraints in sentence comprehension. *Artificial Intelligence*, 46:217–258, 1990.

[99] A. Fagg, D. Lotspeich and G. Bekey. A reinforcement-learning approach to reactive control policy design for autonomous robots. In *Proceedings of the IEEE Conference on Robotics and Automation*, 1994.

[100] D. W. Fausett. Strictly local backpropagation. In *Proceedings of the International Joint Conference on Neural Networks*, pages 125–130, 1990.

[101] L. Fausett. *Fundamentals of Neural Networks*. Prentice-Hall, Englewood Cliffs, NJ, 1994.

[102] J. A. Feldman and D. H. Ballard. Connectionist models and their properties. *Cognitive Science*, 6(3), 1992.

[103] D. Ferster and N. Spruston. Cracking the neuronal code. *Science*, 270:756–757, 1995.

[104] J. Flavell. *The Development Psychology of Jean Piaget*. Van Nostrand Reinhold, New York, 1963.

[105] J. A. Fodor and B. P. McLaughlin. Connectionism and the problem of systematicity—why Smolensky's solution doesn't work. *Cognition*, (35):183–204, 1990.

[106] J. A. Fodor and Z. W. Pylyshyn. Connectionism and cognitive architecture: A critical analysis. *Cognition*, (28):3–71, 1988.

[107] J. A. Fodor. *The Modularity of Mind: An Essay on Faculty Psychology*. Bradford Books/MIT Press, Cambridge, MA, 1983.

[108] J. A. Fodor. The folly of simulation. In P. Baumgartner and S. Payr, editors, *Speaking Minds: Interviews with Twenty Eminent Cognitive Scientists*, pages 84–100, Princeton University Press, Princeton, NJ, 1995.

[109] P. Földiák. Adaptive network for optimal linear feature extraction. In *Proceedings of the International Joint Conference on Neural Networks, IJCNN-89*, pages 401–405, IEEE Press, New York, June 1989.

[110] S. Franklin. *Artificial Minds*. Bradford Books/MIT Press, Cambridge, MA, 1995.

[111] S. Franklin and M. Garzon. Neural computability. In O. Omidvar, editor, *Progress in Neural Networks*, volume 1, Ablex, NJ, 1990.

[112] W. J. Freeman and B. W. van Dijk. Spatial patterns of visual cortical fast EEG during conditioned reflex in a rhesus monkey. *Brain Research*, 422:267–276, 1987.

[113] K. S. Fu, R. C. Gonzalez and C. S. G. Lee. *Robotics: Control, Sensing, Vision and Intelligence*. McGraw-Hill, New York, 1987.

[114] K. S. Fu, R. C. Gonzalez and C. S.G. Lee. *Robotics: Control, Sensing, Vision and Intelligence*. McGraw-Hill, Singapore, 1987.

[115] Berg G. A connectionist parser with recursive sentence structure and lexical disambiguation. In *Proceedings of the Tenth National Conference on Artificial Intelligence*, pages 32–37, MIT Press, Cambridge, MA, 1992.

[116] G. Gallistel. *The Organization of Learning*. MIT Press, Cambridge, MA, 1990.

[117] D. M. Gavrila and F. C. A. Groen. 3D object recognition from 2D images using geometric hashing. *Pattern Recognition Letters*, 13:263–278, 1992.

[118] A. Gelperin, J. J. Hopfield and D. W. Tank. The logic of Limax learning. In A. I. Selverston, editor, *Model Neural Networks and Behavior*, pages 314–339, Plenum, New York, 1985.

[119] P. A. Getting. Reconstruction of small neural networks. In C. Koch and I. Segev, editors, *Methods in Neuronal Modeling: From Synapses to Networks*, pages 319–337, MIT Press, Cambridge, MA, 1989.

[120] J. Gibson. *The Ecological Approach to Visual Perception*. Houghton Mifflin, Boston, MA, 1979.

[121] J. J. Gibson. *The perception of the visual world*. Houghton Mifflin, Boston, MA, 1950.

[122] R. Gill-Carey. Modelling the consciousness. Master's thesis, King's College, London, 1994. Unpublished.

[123] G. W. Cottrell and K. Plunkett. Acquiring the mapping from meaning to sounds. *Connection Science*, 6:379–412, 1994.

[124] R. F. Hadley. Systematicity revisited: Reply to Christiansen and Chater and Niklasson and van Gelder. *Mind and Language*, 9(4), 1994.

[125] R. F. Hadley and V. C. Cardei. Acquisition of the active–passive distinction from sparse input and no error feedback. *Technical Report CSS-IS TR97-01*, School of Computing Science, Simon Frazer University, Burnaby, BC, 1997.

[126] R. F. Hadley. Compositionality and systematicity in connectionist language learning. In *Proceedings of the 14th Annual Conference of the Cognitive Science Society*, pages 659–664. Erlbaum, Hillsdale, NJ, 1992.

[127] G. S. Halford. Creativity and the capacity for representation: Why are humans so creative? *AISB Quarterly*, 85:32–41, 1993.

[128] K. Hammond. *Case-Based Planning*. Academic, San Diego, CA, 1989.

[129] S. Harnad. The symbol grounding problem. *Physica D*, 42:335–346, 1990.

[130] R. D. Hawkins. Interneurons involved in mediation and modulation of gill-withdrawal reflex in Aplysia. III. Identified facilitating neurons increase Ca^{2+} current in sensory neurons. *Journal of Neurophysiology*, 45:327–339, 1981.

[131] R. D. Hawkins, V. F. Castellucci and E. R. Kandel. Interneurons involved in mediation and modulation of gill-withdrawal reflex in Aplysia. I. Identification and characterization. *Journal of Neurophysiology*, 45:304–314, 1981.

[132] R. D. Hawkins, V. F. Castellucci and E. R. Kandel. Interneurons involved in mediation and modulation of gill-withdrawal reflex in Aplysia. II. Identified neurons produce heterosynaptic facilitation contributing to behavioral sensitization. *Journal of Neurophysiology*, 45:315–326, 1981.

[133] R. D. Hawkins and E. R. Kandel. Is there a cell biological alphabet for simple forms of learning? *Psychological Review*, 91:375–391, 1984.

[134] G. Hayes and J. Demiris. A robot controller using learning by imitation. In *Proceedings of the Second International Symposium on Intelligent Robotic Systems*, pages 198–204, 1994.

[135] D. O. Hebb. *The Organization of Behavior*. Wiley, New York, 1949.

[136] E. Hecht and A. Zajac. *Optics*. Addison-Wesley, Reading, MA, 1974.

[137] R. Hecht-Nielsen. Counterpropagation networks. *Applied Optics*, 26:4979–4984, 1987.

[138] J. Heemskerk. *Neurocomputers for Brain-Style Processing: Design, Implementation and Application*. PhD thesis, Department of Experimental and Theoretical Psychology, The Netherlands, 1995.

[139] J. Hendler. Marker passing and microfeatures. In *Proceedings of the Tenth International Joint Conference on Artificial Intelligence*, pages 151–154, Morgan Kaufmann, San Mateo, CA, 1987.

[140] J. Hertz, A. Krogh and R. G. Palmer. *Introduction to the Theory of Neural Computation*. Addison-Wesley, Reading, MA, 1991.

[141] G. E. Hinton, J. L. McClelland and D. E. Rumelhart. Distributed representations. In D. E. Rumelhart and J. L. McClelland, editors, *Parallel Distributed Processing*, volume 1, pages 77–109. MIT Press, Cambridge, MA, 1986.

[142] G. E. Hinton and T. Shallice. Lesioning an attractor network: Investigations of acquired dyslexia. *Psychological Review*, 98(1):74–95, 1991.

[143] G. E. Hinton. Mapping part–whole hierarchies into connectionist networks. *Artificial Intelligence*, (46):47–75, 1990.

[144] S. Hölldobler. CHCL—a connectionist inference system for Horn logic based on the connection method. *Technical Report TR-90-042*, International Computer Science Institute, Berkeley, CA, 1990.

[145] J. J. Hopfield. Neural computations and neural systems. In R. M. Cotterill, editor, *Computer Simulation in Brain Science*, pages 113–120, Cambridge University Press, Cambridge, 1988.

[146] P. V. Horne. The nature of imagery. *Consciousness and Cognition*, 2:58–82, 1993.

[147] K. Hornik, M. Stinchcombe and H. White. Universal approximation of an unknown mapping and its derivatives using multilayer feedforward networks. *Neural Networks*, 3:551–560, 1990.

[148] I. Aleksander. *An Introduction to Neural Computing*. Thomson, London, 1995.

[149] I. Aleksander. *The Impossible Mind of MAGNUS*. Imperial College Press, London, 1996.

[150] D. J. Foss and H. S. Cairns. Some effects of memory limitation upon sentence comprehension and recall. *Journal of Verbal Learning and Verbal Behavior*, 9:541–547, 1970.

[151] J. Weckerly and J. L. Elman. A PDP approach to processing center-embedded sentences. In *Proceedings of the 14th Annual Conference of the Cognitive Science Society*, pages 414–419, Erlbaum, Hillsdale, NJ, 1992.

[152] W. James. *Psychology: A briefer course*. Harper, New York, 1892, reprinted 1961.

[153] M. I. Jordan. Attractor dynamics and parallelism in a connectionist sequential machine. In *Proceedings of the Eighth Conference of the Cognitive Science Society*, pages 531–546, 1986.

[154] W. Kabat and A. Wojcik. Automated synthesis of combinatorial logic using theorem proving techniques. *IEEE Transactions on Computing*, C-34:610–628, 1985.

[155] E. R. Kandel. Small systems of neurons. *Scientific American*, 241:61–70, 1979.

[156] I. Kant. *A Critique of Pure Reason*. MacMillan, New York, 1953.

[157] A. Karmiloff-Smith. *Beyond Modularity: A Developmental Perspective on Cognitive Science*. MIT Press, Cambridge, MA, 1992.

[158] N. Kasabov and S. Shishkov. A connectionist production system with partial match and its use for approximate reasoning. *Connection Science*, 5(3 and 4):275–306, 1993.

[159] F. Keijzer and J. Heemskerk. Proximal and distal descriptions of adaptive behavior. Unpublished manuscript, 1995.

[160] J. A. S. Kelso, S. L. Bressler, S. Buchanan, G. C. DeGuzman, M. Ding, A. Fuchs, and T. Holroyd. A phase transition in human brain and behavior. *Physics Letters*, 169A:134–144, 1992.

[161] J. Klahr. *Production System Models of Learning and Development*. MIT Press, Cambridge, MA, 1989.

[162] D. Kleinfield and H. Sompolinsky. Synaptic modelling. In C. Koch and I. Segev, editors, *Methods in Neuronal Modeling: From Synapses to Networks*, pages 214–263, MIT Press, Cambridge, MA, 1989.

[163] A. H. Klopf, J. S. Mogan and S.E Weaver. Modeling nervous system function with a hierarchical network of control systems that learn. In J.-H. Meyer, H. L. Roitblat and S. W. Wilson, editors, *From Animals to Animats 2: Proceedings of the Second International Conference on Simulation of Adative Behavior*, pages 254–261, Bradford Books/MIT Press, Cambridge, MA, 1992.

[164] C. Koch and I. Segev, editors. *Methods in Neuronal Modeling: From Synapses to Networks*. MIT Press, Cambridge, MA, 1989.

[165] S. W. Kuffler, J. G. Nicholls and A. R. Martin. *From Neuron to Brain: A Cellular Approach to the Function of the Nervous System*. Sinauer, Sunderland, MA, second edition, 1984.

[166] Y. Kuniyosi, M. Inaba and H. Inoue. Learning by watching: Extracting reusable task knowledge from visual observation of human performance. *IEEE Transactions on Robotics and Automation*, 10(6):799–822, 1994.

[167] J. L. Elman. Distributed representations, simple recurrent networks, and grammatical structure. *Machine Learning*, 7:195–225, 1991.

[168] J. L. McClelland and A. H. Kawamoto. Mechanisms of sentence processing: Assigning roles to constituents. In J. L. McClelland and D. E. Rumelhart, editors, *Parallel Distributed Processing: Explorations in the Microstructure of Cognition, Volume 2: Psychological and Biological Models*, pages 272–325. MIT Press, Cambridge, MA, 1986.

[169] P. Ladefoged. *A Course in Phonetics*. Harcourt Brace, 1993.

[170] T. Lange and M. G. Dyer. High-level inferencing in a connectionist network. *Technical Report UCLA-AI-89-12*, UCLA, Los Angeles, 1989.

[171] D. N. Lee. The optic flow field: the foundation of vision. *Philosophical Transactions of the Royal Society of London B*, 290:169–179, 1980.

[172] G. Lee, M. Flowers and M. G. Dyer. Learning distributed representations for conceptual knowledge and their application to script-based story processing. *Connection Science*, 2(4):313–345, 1990.

[173] I. B. Levitan and L. K. Kaczmarek. *The Neuron: Cell and Molecular Biology*. Oxford University Press, Oxford, 1991.

[174] W. Li, N. H. Packard and C. G. Langton. Transition phenomena in cellular automata rule space. *Physica D*, 45:77–94, 1990.

[175] B. Libet. Brain stimulation in the study of neuronal functions for conscious sensory experience. *Human Neurobiology*, 1:235–242, 1982.

[176] B. Libet, W. W. Alberts, E. W. Wright, D. L. Delattre, G. Levin and B. Feinstein. Production of threshold levels of conscious sensation by electrical stimulation of the human somato-sensory cortex. *Journal of Neurophysiology*, 27:546–578, 1964.

[177] R. Linsker. Self-organization in a perceptual network. *IEEE Computer*, 21(3):105–117, March 1988.

[178] R. R. Llinas. The intrinsic electrophysiological properties of mammalian neurons: insights into central nervous system function. *Science*, 242:1654–1664, 1988.

[179] D. Lloyd. *Simple Minds*. MIT Press, Cambridge, MA, 1989.

[180] J. Loeb. *Forced Movements, Tropisms and Animal Conduct*. Lippincott, Philadelphia, PA, 1918.

[181] S. Löwel and W. Singer. Columnar specificity of intrinsic horizontal connections in the visual cortex of strabismic cats. *European Journal of Neuroscience*, page 52, 1991.

[182] W. Maas. Networks of spiking neurons: The third generation of neural network models. *Technical Report NC-TR-96-045*, Technische Universitaet Graz, 1996.

[183] R. J. MacGregor. *Neural and Brain Modelling*. Academic, London, 1987.

[184] D. J. C. MacKay. Information-based objective functions for active data selection. *Neural Computation*, 4:590–604, 1992.

[185] A. Marcel. Conscious and preconscious recognition of polysemous words. In R. S. Nickerson, editor, *Attention and Performance*, volume VIII. Erlbaum, Hillsdale, NJ, 1980.

[186] Z. Markov. A tool for building connectionist-like networks based on term unification. In *Proceedings of the Processing Declarative Knowledge International Workshop*, pages 199–213, 1991.

[187] D. Marr. *Vision*. Freeman, New York, 1982.

[188] D. Marr. *Vision: A Computational Investigation into the Human Representation and Processing of Visual Information*. Freeman, New York, 1982.

[189] M. Mataric and D. Cliff. Challenges in evolving controllers for physical robots. *Robotics and Autonomous Systems*, 1996. In press.

[190] J. H. R. Maunsell and J. Gibson. Visual response latencies in the striate cortex of the macaque monkey. *Journal of Neurophysiology*, 68:1332–1344, 1992.

[191] J. McCarthy. Circumscription: A form of non-monotonic reasoning. *Artificial Intelligence*, 13:27–39, 1980.

[192] J. L. McClelland and D. E. Rumelhart. An interactive activation model of effects in letter perception. *Psychological Review*, 88:375–407, 1981.

[193] J. L. McClelland and D. E. Rumelhart. Distributed memory and the representation of general and specific information. *Journal of Experimental Psychology General*, 114(2):159–188, 1985.

[194] W. S. McCulloch and W. Pitts. A logical calculus of ideas immanent in nervous activity. *Bulletin of Mathematical Biophysics*, 5:115–133, 1943.

[195] L. Meeden, G. McGraw and D. Blank. Emergent control and planning in an autonomous vehicle. In *Proceedings of the Fifteenth Annual Conference of the Cognitive Science Society*, pages 735–740, 1993.

[196] M. Minsky. A framework for representing knowledge. In P. H. Winston, editor, *The Psychology of Computer Vision*. McGraw-Hill, New York, 1975.

[197] M. L. Minsky and S. Papert. *Perceptrons:An Introduction to Computational Geometry: Expanded Edition*. MIT Press, Cambridge, MA, 1988.

[198] M. Mitchell, P. T. Hraber and J. P. Crutchfield. Revisiting the edge of chaos: Evolving cellular automata to perform computations. *Complex Systems*, 1993.

[199] E. Mjolness, E. Gindi and P. Anandan. Optimization in model matching. *Neural Computation*, 1:218–229, 1989.

[200] R. C. Moore. Semantical considerations in non-monotonic logic. *Artificial Intelligence*, 25:75–94, 1985.

[201] J. M. Murre. Neurosimulators. In M. A. Arbib, editor, *Handbook of Brain Theory and Neural Networks*, pages 634–639, MIT Press, Cambridge, MA, 1995.

[202] A. N. Jain. Parsing complex sentences with structured connectionist networks. *Neural Computation*, 3:110–120, 1991.

[203] T. Nagel. What is it like to be a bat? *Philosophical Reviews*, 83:435–450, 1974.

[204] K. S. Narendra and A. M. Annaswamy. *Stable Adaptive Systems*. Prentice-Hall, Englewood Cliffs, NJ, 1989.

[205] K. S. Narendra and K. Parthasarathy. Identification and control of dynamical systems using neural networks. *IEEE Transactions on Neural Networks*, 1(1):4–27, March 1990.

[206] U. Nehmzow. Self-organization and self-learning robot control. In *Proceedings of the IEE Seminar on Self-Learning Robots*, Digest No: 96/026, 1996.

[207] U. Nehmzow and B. McGonigle. Achieving rapid adaptations in robots by means of external tuition. In D. T. Cliff *et al*, editors, *From Animals to Animats 3: Proceedings of the Third International Conference on Simulation of Adaptive Behaviour*, pages 87–96, Bradford Books/MIT Press, Cambridge, MA, 1994.

[208] H. Nema. Phonotactis and sonority (in Japanese). *Senshu Journal of Foreign Language and Education*, pages 65–94, 1994.

[209] J. Von Neumann. In A. W. Burks, editor, *Theory of Self-Reproducing Automata*. University of Illinois Press, Urbana, IL, 1966.

[210] A. Newell. Physical symbol systems. *Cognitive Science*, 4:135–183, 1980.

[211] A. Newell. The symbol level and the knowledge level. In Z. W. Pylyshyn and W. Demopoulos, editors, *Meaning and Cognitive Structure*, pages 31–39. Ablex, Norwood, NJ, 1986.

[212] A. Newell and H. Simon. *Human Problem Solving*. Prentice-Hall, Englewood Cliffs, NJ, 1972.

[213] L. Niklasson and T. van Gelder. Can connectionist models exhibit non-classical structure sensitivity? In *Proceedings of the Cognitive Science Society*, pages 664–669. Erlbaum, Hillsdale, NJ, 1994.

[214] L. F. Niklasson. Structure sensitivity in connectionist models. In *Proceedings of the 1993 Connectionist Models Summer School*, pages 162–169, Erlbaum, Hillsdale, NJ, 1993.

[215] L. F. Niklasson and N. E. Sharkey. Connectionism and the issues of compositionality and systematicity. In R. Trappl, editor, *Cybernetics and Systems Research*, pages 59–71. World Scientific, Singapore, 1992.

[216] M. Nilsson. Toward conscious robots. In *Proceedings of the IEE Seminar on Self-Learning Robots*, 1996. Digest No: 96/026.

[217] M. Nilsson and J. Ojala. Self-awareness in reinforcement learning of snake-like robot locomotion. In *Proceedings IASTED 95*, 1995.

[218] N. Nilsson. Shakey the robot. *Technical Note 323*, SRI International, Menlo Park, CA, 1984.

[219] S. Nolfi and D. Parisi. Evolving non-trivial behaviors on real robots: an autonomous robot that picks up objects. In *Proceedings of the Fourth Congress of the Italian Association of Artificial Intelligence*, pages 243–254, 1995.

[220] E. Oja. A simplified neuron model as a principal component analyser. *Journal of Mathematical Biology*, 15:267–273, 1982.

[221] E. Oja and J. Karhunen. On stochastic approximation of the eigenvectors and eigenvalues of the expectation of a random matrix. *Journal of Mathematical Analysis and Applications*, 106:69–84, 1985.

[222] R. D. Orpwood. Basic module for an adaptive control system based on neurone information processing. *Journal of Biomedical Engineering*, 10:201–205, 1988.

[223] P. Munro, C. Cosic and M. Tabasko. A network for encoding, decoding and translating locative prepositions. *Connection Science*, 3:225–240, 1991.

[224] N. H. Packard. Adaptation at the edge of chaos. In J. A. S. Kelso, A. J. Mandell and M. F. Schlesinger, editors, *Dynamic Patterns in Complex Systems*. World Scientific, Singapore, 1988.

[225] S. E. Palmer. Fundamental aspects of cognitive representation. In E. Rosch and B. B. Lloyd, editors, *Cognition and Categorization*. Erlbaum, Hillsdale, NJ, 1978.

[226] I. Pavlov. *Conditioned Reflexes*. Oxford University Press, London, 1927.

[227] J. Pearl. *Probabilistic Reasoning in Intelligent Systems*. Morgan Kaufmann, San Mateo, CA, 1988.

[228] R. Penrose. *The Emperor's New Mind*. Oxford University Press, Oxford, 1989.

[229] D. Perlis. On the consistency of commonsense reasoning. *Computational Intelligence*, 2:180–190, 1986.

[230] S. Phillips. Strong systematicity within connectionism: The tensor-recurrent network. In *Proceedings of the Sixteenth Annual Conference of the Cognitive Science Society*, pages 723–727, 1994.

[231] J. Piaget. *The Psychology of Intelligence*. Routledge and Kegan Paul, London, 1950.

[232] J. Piaget. *The Origins of Intelligence*. International Universities Press, New York, 1952.

[233] S. Pinker and A. Prince. Language and connectionism. In S. Pinker and J. Mehler, editors, *Connections and Symbols*, pages 73–193. MIT Press, Cambridge, MA, 1988. Reprinted from *Cognition*, vol. 28, 1988.

[234] T. Plate. *Distributed Representations and Nested Compositional Structure*. PhD thesis, University of Toronto, 1994.

[235] D. C. Plaut and T. Shallice. Deep dyslexia: A case study of connectionist neuropsychology. *Cognitive Neuropsychology*, 10(5):377–500, 1993.

[236] M. D. Plumbley. On information theory and unsupervised neural networks. *Technical Report CUED/F-INFENG/TR.78*, Cambridge University Engineering Department, 1991.

[237] M. D. Plumbley. Efficient information transfer and anti-Hebbian neural networks. *Neural Networks*, 6:823–833, 1993.

[238] M. D. Plumbley. A Hebbian/anti-Hebbian network which optimizes information capacity by orthonormalizing the principal subspace. In *Proceedings of the IEE Artificial Neural Networks Conference, ANN-93, Brighton*, pages 86–90, 1993.

[239] M. D. Plumbley. Lyapunov functions for convergence of principal component algorithms. *Neural Networks*, 8:11–23, 1995.

[240] M. D. Plumbley and F. Fallside. An information-theoretic approach to unsupervised connectionist models. In D. Touretzky, G. Hinton and T. Sejnowski, editors, *Proceedings of the 1988 Connectionist Models Summer School*, pages 239–245. Morgan Kaufmann, San Mateo, CA, 1988.

[241] J. B. Pollack. Recursive auto-associative memory—devising compositional distributed representations. In *Proceedings of the Tenth Annual Conference of the Cognitive Science Society*, pages 33–39, 1988.

[242] D. A. Pomerleau. *Neural Network Perception for Mobile Robot Guidance*. Kluwer, Deventer, 1993.

[243] R. F. Port and T. van Gelder. MIT Press, Cambridge, MA, 1995.

[244] L. Y. Pratt, J. Mostow and C. A. Kamm. Direct transfer of learned information among neural networks. In *Proceedings of AAAI 91*, pages 584–589, 1991.

[245] Z. Pylyshyn. Computation and cognition: Issues in the foundations of cognitive science. *Behavioral and Brain Sciences*, 3:111–132, 1980.

[246] Z. W. Pylyshyn. *Computation and Cognition: Toward a Foundation for Cognitive Science*. MIT Press, Cambridge, MA, 1984.

[247] J. R. Quinlan. Induction of decision trees. *Machine Learning*, 1:81–106, 1986.

[248] L. F. R. Karen. Identification of topical entities in discourse: A connectionist approach to attentional mechanisms in language. *Connection Science*, 2:103–122, 1990.

[249] J. R. Lucas. A view of one's own. *Philosophical Transactions of the Royal Society of London*, A(394):147–152, 1994.

[250] R. Miikkulainen. *Subsymbolic Natural Language Processing: An Integrated Model of Scripts, Lexicon and Memory*. MIT Press, Cambridge, MA, 1993.

[251] R. Miikkulainen. Subsymbolic case-role analysis of sentences with embedded clauses. *Cognitive Science*, 20:47–73, 1996.

[252] R. Penrose. *Shadows of the Mind*. Oxford University Press, Oxford, 1994.

[253] W. Rall. Cable theory for dendritic neurons. In C. Koch and I. Segev, editors, *Methods in Neuronal Modeling: From Synapses to Networks*, pages 315–389, MIT Press, Cambridge, MA, 1989.

[254] W. Reichardt. Movement perception in insects. In W. Reichardt, editor, *Processing of Optical Data by Organisms and Machines*, pages 465–493, Academic, New York, 1969.

[255] R. Reiter. A logic for default reasoning. *Artificial Intelligence*, 13:81–132, 1980.

[256] R. A. Rescorla and A. R. Wagner. A theory of Pavlovian conditioning: the effectiveness of reinforcement and non-reinforcement. In A. H. Black and W. F. Prokasy, editors, *Classical Conditioning II: Current Research and Theory*, pages 64–69, Appleton-Century-Crofts, New York, 1972.

[257] C. Riesback and R. Schank. *Inside Case-Based Reasoning*, Erlbaum, Hillsdale, NJ, 1989.

[258] H. J. Ritter, T. M. Martinetz and K. J. Schulten. Topology-conserving maps for learning visuo-motor-coordination. 2:159–168, 1989.

[259] J. A. Robinson. A machine-oriented logic based on the resolution principle. *Journal of the Association for Computing Machinery*, 12:23–41, 1965.

[260] E. Rolls. The representation and storage of information in neural networks in the primate cerebral cortex and hippocampus. In R. Durbin, C. Miall and G. Mitchison, editors, *The Computing Neuron*, pages 389–413, Addison-Wesley, Reading, MA, 1989.

[261] R. Rosenfeld and D. Touretzky. Coarse coded symbol memories and their properties. *Complex Systems*, 2:463–484, 1988.

[262] D. E. Rumelhart and J. L. McClelland. Levels indeed! *Journal of Experimental Psychology General*, 114(2):193–197, 1985.

[263] D. E. Rumelhart, G. E. Hinton and R. J. Williams. Learning internal representations by back-propagating errors. *Nature*, (323):533–536, 1986.

[264] M. S. Huang. A developmental study of children's comprehension of embedded sentences with and without semantic constraints. *Journal of Psychology*, 114:51–56, 1983.

[265] W. S. Stolz. A study of the ability to decode grammatically novel sentences. *Journal of Verbal Learning and Verbal Behavior*, 6:867–873, 1967.

[266] D. S. Touretzky. Connectionism and compositional semantics. In J. A. Barnden and J. B. Pollack, editors, *High-Level Connectionist Models, Volume 1 of Advances in Connectionist and Neural Computation Theory*, pages 17–31. Ablex, Norwood, NJ, 1991.

[267] T. D. Sanger. An optimality principle for unsupervised learning. In D. S. Touretzky, editor, *Advances in Neural Information Processing Systems 1*, pages 11–19. Morgan Kaufmann, San Mateo, CA, 1989.

[268] R. A. Satterlie. Reciprocal inhibition and postinhibitory rebound produce reverberation in a locomotor pattern generator. *Science*, 229:402–404, 1985.

[269] R. C. Schank. The role of memory in language processing. In C. N. Cofer, editor, *The Structure of Human Memory*, pages 162–189. Freeman, San Francisco, CA, 1976.

[270] N. A. Schmajuk. Conditioning. In M. A. Arbib, editor, *Handbook of Brain Theory and Neural Networks*, pages 238–243, MIT Press, Cambridge, MA, 1995.

[271] B. Schwartz. *Psychology of Learning and Behavior (Second Edition)*. Norton, New York, 1984.

[272] T. W. Scutt and R. I. Damper. Computational modelling of learning and behaviour in small neuronal systems. In *Proceedings of International Joint Conference on Neural Networks (Singapore, 1991)*, pages 430–435, 1991.

[273] I. Segev, J. W. Fleshman and R. E. Burke. Compartmental models of complex neurons. In C. Koch and I. Segev, editors, *Methods in Neuronal Modeling: From Synapses to Networks*, pages 319–337, MIT Press, Cambridge, MA, 1989.

[274] I. Segev and W. Rall. Computational study of an excitable dendritic spine. *Journal of Neurophysiology*, 60:499–523, 1988.

[275] T. J. Sejnowski, C. Koch and P. S. Churchland. Computational neuroscience. *Science*, 241:1299–1306, 1988.

[276] A. I. Selverston. A consideration of invertebrate central pattern generators as computational data bases. *Neural Networks*, 1:109–117, 1988.

[277] A. I. Selverston and M. Moulins. Oscillatory neural networks. *Review of Physiology*, 47:29–48, 1985.

[278] C. E. Shannon. A mathematical theory of communication. *Bell System Technical Journal*, 27:379–423, 623–656, 1948.

[279] C. E. Shannon. Communication in the presence of noise. *Proceedings of the IRE*, 37:10–21, 1949.

[280] N. Sharkey, J. Heemskerk and J. Neary. Subsuming behaviors in neural network controllers. In *Proceedings of Robolearn 96*, 1996.

[281] N. E. Sharkey. The ghost in the hybrid—a study of uniquely connectionist representations. *AISB Quarterly*, (79):10–16, 1992.

[282] N. E. Sharkey and S. A. Jackson. Three horns of the representational trilemma. In V. Honavar, editor, *Artificial Intelligence and Neural Networks: Steps Towards Principled Integration*, pages 155–189. Academic, Cambridge, MA, 1994.

[283] N. E. Sharkey. Functional compositionality and soft preference rules. In R. Linggard, D. J. Myers and C. Nightingale, editors, *Neural Networks for Vision, Speech and Natural Language*, pages 235–255. Chapman and Hall, London, 1992.

[284] L. Shastri and V. Ajjanagadde. A connectionist system for rule based reasoning with multi-place predicates and variables. *Technical Report MS-CIS-89-06*, University of Pennsylvania, Philadelphia, PA, 1989.

[285] L. Shastri and V. Ajjanagadde. From simple associations to systematic reasoning: A connectionist representation of rules, variables and dynamic bindings. *Technical Report MS-CIS-90-05*, University of Pennsylvania, Philadelphia, PA, 1990.

[286] G. M. Shepherd. *Neurobiology (Third Edition)*. Oxford University Press, New York, 1994.

[287] B. Skinner. *Science and Human Behavior*. MacMillan, New York, 1953.

[288] E. Smith, C. Langston and R. Nisbett. The case for rules in reasoning. *Cognitive Science*, 16:1–40, 1992.

[289] E. Smith and D. Medin. *Categories and Concepts*. Harvard University Press, Cambridge, MA, 1981.

[290] P. Smolensky. On the proper treatment of connectionism. *Behavioral and Brain Sciences*, 11:1–74, 1988.

[291] P. Smolensky and O. Brousse. Virtual memories and massive generalisation in connectionist combinatorial learning. In *Proceedings of the Tenth Annual Conference of the Cognitive Science Society*, pages 380–387, 1989.

[292] P. Smolensky. Tensor product variable binding and the representation of symbolic structures in connectionist systems. *Artificial Intelligence*, 46:159–216, 1990.

[293] J. Staddon. *Behaviorism*. Duckworth, London, 1993.

[294] A. Stolke. A connectionist model of unification. *Technical Report TR-89-032*, International Computer Science Institute, Berkeley, CA, 1989.

[295] R. Sun. A discrete neural network model for conceptual representation and reasoning. In *Proceedings of the 11th Conference of the Cognitive Science Society*, pages 916–923, Erlbaum, Hillsdale, NJ, 1989.

[296] R. Sun. On variable binding in connectionist networks. *Connection Science*, 4:93–124, 1992.

[297] R. Sun. *Integrating Rules and Connectionism for Robust Commonsense Reasoning*. Wiley, New York, 1994.

[298] R. Sun. A new approach towards modelling causality in commonsense reasoning. *International Journal of Intelligent Systems*, 10:581–616, 1995.

[299] R. Sun. Robust reasoning: Integrating rule-based and similarity-based reasoning. *Artificial Intelligence*, 75:241–295, 1995.

[300] R. Sun and L. A. Bookman. *Computational Architectures Integrating Neural and Symbolic Processes: A Perspective on the State of the Art*. Kluwer, Boston, MA, 1994.

[301] R. Sun and D. Waltz. Neurally inspired massively parallel model of rule-based reasoning. In B. Soucek, editor, *Neural and Intelligent System Integration*, pages 341–381. Wiley, New York, 1991.

[302] R. S. Sutton. Reinforcement learning architectures. In J.-A. Meyer and S. W. Wilson, editors, *From Animals to Animats: Proceedings of the First International Conference on Simulation of Adaptive Behaviour*, pages 288–296, Bradford Books/MIT Press, Cambridge, MA, 1991.

[303] R. S. Sutton and A. G. Barto. Towards a modern theory of adaptive networks: expectation and prediction. *Psychological Review*, 88:135–170, 1981.

[304] T. Nagel. *The View From Nowhere*. Oxford University Press, Oxford, 1982.

[305] J. G. Taylor. A silicon model of vertebrate retinal processing. *Neural Networks*, 3:171–178, 1990.

[306] J. G. Taylor. Can neural networks ever be made to think? *Neural Network World*, 1:4–12, 1991.

[307] J. G. Taylor. Towards a neural network model of the mind. *Neural Network World*, 2:797–812, 1992.

[308] J. G. Taylor. A global gating model of attention and consciousness. In M. Oaksford and G. Brown, editors, *Neurodynamics and Psychology*, pages 31–52. Academic, New York, 1993.

[309] J. G. Taylor. Modelling the mind. In G. M. Orchard, editor, *Neural Computing Research and Applications*, volume 1, pages 1–20. Institute of Physics Publishing, Bristol, 1993.

[310] J. G. Taylor. *The Promise of Neural Networks*. Springer, Berlin, 1993.

[311] J. G. Taylor. Breakthrough to awareness. *Biological Cybernetics*, 1996. In Press.

[312] J. G. Taylor. *The Emergent Mind*. MIT Press, Cambridge, MA, 1996. In Press.

[313] R. F. Thompson. The neurobiology of learning and memory. *Science*, 233:941–947, 1986.

[314] H. Tomabechi and H. Kitano. Beyond PDP: The frequency modulation neural network architecture. In *Proceedings of the International Joint Conference on Artificial Intelligence*, pages 186–192, 1989.

[315] D. Touretzky and G. Hinton. Symbols among the neurons. In *Proceedings of the Ninth International Joint Conference on Artificial Intelligence*, pages 238–243, Morgan Kaufmann, San Mateo, CA, 1985.

[316] D. S. Touretzky and G. E. Hinton. A distributed connectionist production system. *Cognitive Science*, 12(3):423–466, 1988.

[317] C. Touzet. Neural implementations of immediate reinforcement learning for an obstacle avoidance behavior. *Technical Report nm.94.6*, Laboratoire d'Etude et de Recherche en Informatique, Nimes, 1994.

[318] H. C. Tuckwell. *Introduction to Theoretical Neurobiology: Vol. 1—Linear Cable Theory and Dendritic Structure*. Cambridge University Press, New York, 1988.

[319] A. Tversky. Features of similarity. *Psychological Review*, 84(4):327–352, 1977.

[320] J. Von Uexkuell. *Umwelt und Innenwelt der Tiere*. Springer, Berlin, 1921.

[321] D. I. Uzunov. *Theory of Critical Phenomena: Mean Field, Fluctuations and Renormalization*. World Scientific, Singapore, 1993.

[322] P. van der Smagt. Minimisation methods for training feed-forward networks. *Neural Networks*, 7(1):1–11, 1994.

[323] P. van der Smagt, A. Jansen and F. Groen. Interpolative robot control with the nested network approach. In *IEEE International Symposium on Intelligent Control (Glasgow, Scotland, August 1992)*, pages 475–480. IEEE, New York, 1992.

[324] P. van der Smagt. *Visual Robot Arm Guidance using Neural Networks*. PhD thesis, Department of Computer Systems, University of Amsterdam, March 1995.

[325] T. van Gelder. Compositionality: A connectionist variation on a classical theme. *Cognitive Science*, 14:335–364, 1990.

[326] T. van Gelder. What is the 'D' in 'PDP'? A survey of the concept of distribution. In W. Ramsey, S. Stich and D. E. Rumelhart, editors, *Philosophy and Connectionist Theory*, pages 33–60. Erlbaum, Hillsdale, NJ, 1991.

[327] F. Varela, E. Thompson and E. Rosch. *The Embodied Mind: Cognitive Science and Human Experience*. MIT Press, Cambridge, MA, 1991.

[328] P. Verschure, B. Krose and R. Pfeifer. Distributed adaptive control: The self organization of structured behavior. *Robotics and Autonomous Systems*, 9(2):181–196, 1992.

[329] C. von der Malsberg and E. Bienenstock. Statistical coding and short term plasticity: A scheme for knowledge representation in the brain. In E. Bienenstock, F. Fogelman and G. Weisbuch, editors, *Disordered Systems and Biological Organisation*, pages 247–272. Springer, Berlin, 1986.

[330] G. Walter. *The Living Brain*. Norton, New York, 1953.

[331] S. Watanabe. *Pattern Recognition: Human and Mechanical*. Wiley, New York, 1985.

[332] R. J. Williams. Feature discovery through error-correction learning. *ICS Report 8501*, University of California, San Diego, CA, 1985.

[333] S. Wilson. Explore/exploit strategies in autonomous learning. In *Proceedings of the Workshop on Learning in Robots and Animals, AISB 96*, 1996.

[334] R. Wolf, A. Voss, S. Hein and M. Heisenberg. Can a fly ride a bicycle? *Philosophical Transactions of the Royal Society of London (B) Biological Sciences*, 337:261–269, 1992.

[335] W. M. Yamada, C. Koch and P. R. Adams. Multiple channels and calcium dynamics. In C. Koch and I. Segev, editors, *Methods in Neuronal Modeling: From Synapses to Networks*, pages ???–???, MIT Press, Cambridge, MA, 1989.

[336] M. P. Young, K. Tanaka and S. Yamane. On oscillating neuronal responses in the visual cortex of the monkey. *Journal of Neurophysiology*, 67:1464–1474, 1992.

[337] L. Zadeh. Fuzzy logic. *Computer*, 21(4):83–93, 1988.

[338] T. Ziemke. Towards autonomous robot control using connectionist infinite state automata. In *Proceedings of the Workshop on Learning in Robots and Animals, AISB 96*, 1996.

[339] T. Sejnowski and C. Rosenberg. Parallel networks that learn to pronounce English text. *Complex Systems*, (1):145–168, 1987.

[340] D. Sanger. Contribution analysis: A technique for assigning responsibilities to hidden units in connectionist networks. *Connection Science*, (1):115–138, 1989.

[341] N. E. Sharkey and S. Jackson. An internal report for connectionists. In R. Sun and L. Bookman, editors, *Computational Architectures Integrating Neural and Symbolic Processes*, pages 223–244. Kluwer, Boston, MA, 1994.

[342] S. J. Hanson and D. J. Burr. What connectionist models learn: Learning and representation in connectionist networks. *Behavioral and Brain Sciences*, (13):471–518, 1990.

[343] J. Hopfield. *Proceedings of the National Academy of Sciences*, 81:3088–3092, 1984.

[344] D. E. Rumelhart, G. E. Hinton and R. J. Williams. *Learning Internal Representations by Error Propagation*, volume 1. MIT Press, Cambridge, MA, 1986.

[345] G. A. Carpenter and S. Grossberg. The art of adaptive pattern recognition by a self-organizing neural network. *Computer*, pages 77–88, March 1988.

[346] J. L. Elman. Finding structure in time. *Cognitive Science*, 14:179–211, 1990.

[347] T. Kohonen. Springer, New York, 1988.

Index

adaptive control, 203
ambiguity in arm posture, 196
associative learning, 226, 227
attractors, 55–58

binning of learning samples, 209
binocular vision, 200

camera
    external vs. internal, 199
    pinhole, 201
case-based reasoning, 105, 108
CCD, 201
centrifugal forces, 198
chained conditioning, 248
chaos, 56–58
classical conditioning, 245–247
closed loop control, 207, 208
component recognition, 202
compositionality, 24, 26–28, 30–40,
    42
conjugate gradient, 205
consciousness, 148–150, 153–156,
    158–160, 162, 165, 166
control
    adaptive, 203
        direct, 203
        indirect, 203
    closed loop, 207, 208
    deterministic, 202
    open loop, 207
    stochastic, 203
controller

ideal, 207
Coriolis forces, 198
correspondence problem, 201

decision hyperplanes, 43, 45, 46
deterministic control process, 202
direct adaptive control, 203
distal systems, 174, 175
dynamic networks, 55–58

elbow up–elbow down ambiguity,
    196
entropy, 73–75
evolutionary learning, 184, 185
external vs. internal camera, 199

friction, 198
fuzzy logic, 99, 103

goal, 207
gravity, 198

habituation, 224–226, 234, 235, 237
hebbian learning, 178, 179
hierarchical cluster analysis, 42, 43
hinton diagrams, 46–49

I-max, 87–89
ideal controller, 207
image acquisition, 202
image description, 202
image segmentation, 202
indirect adaptive control, 203
infomax, 76, 77, 81, 82

information theory, 72, 73
internal vs. external camera, 199
interpolation learning, 205, 206
inverse dynamics, 198

Kohonen networks, 187, 188

learning
   by interpolation, 206
learning samples, 206
   binning, 209
   interpolation, 206
   set of, 196

monocular vision, 200

neural state machine, 156, 158–160,
      162, 165, 166
nucleus reticularis thalami, 142–144

occlusion, 201
open loop control, 207
optic flow, 212

perception, 199–202
physical symbol system, 22–24
pinhole camera, 201
principal components analysis, 77–81
   multidimensional, 82, 83
probabilistic reasoning, 98, 99
proximal systems, 174, 175

recurrent networks, 131, 186, 187
recursive auto-associative memory,
      31–40, 42, 131, 137, 138

reinforcement learning, 182–184
robot
   ambiguity in posture, 196
   perception, 199–202
   vision, 199–202
rule-based reasoning, 95, 98, 104,
      105
rules, 53, 54

sensitization, 224–226
sensory strip, 162, 163
setpoint, 206
sockets, 208
stereoscopic vision, 200
stochastic control process, 203
subsymbolic representations, 121,
      123–125, 127–131, 135–138
symbolic systems, 51–53
systematicity, 24, 26–28, 30–40, 42

tensor products, 28, 30
time-dependent constraint, 216

vision, 199–202
   binocular, 200
   component recognition, 202
   image acquisition, 202
   image description, 202
   image segmentation, 202
   monocular, 200
   occlusion, 201
   stereoscopic, 200
   timing, 208